# All Roads Lead to Deadwood

By Irma H. Klock

Library of Congress catalog card number 78-78323
Manufactured in the United States of America
0-87970-147-1

## ACKNOWLEDGEMENTS

A great many knowledgeable people helped trace the old stage and freight trails. They shared information, maps and photographs. They gave of their time, their hospitality and many offered the use of vehicles in rugged terrain. Several individuals led to valuable sources by suggesting going to a neighbor, a former resident or a certain book or map. And not to be overlooked are the countless area residents who courteously gave directions to find the trail or a station—that often were most elusive.

A special thanks to the library personnel including the librarians at the Phoebe Apperson Hearst Free Library, Lead, the Andrew Carneige Library, Deadwood, the Grace Balloch Library, Spearfish, the Leland Case Library at Black Hills State College, the Sturgis Public Library and the Miles City Public Library.

Near the top of the list are the staff members of the historical societies who were most helpful—the South Dakota Historical Society, the State Historical Society of North Dakota, the Nebraska Historical Society, the personnel at Fort Robinson—part of the Nebraska Historical Society and the Wyoming Historical Department.

With the aid of all these good people it was possible to travel the rainbow trails that arched from all directions to Deadwood, the pot-of-gold. My utmost thanks.

The following people who helped are: Marvin Aga, Raymond Andersen, Melvin Anderson, Albert Auer, Ralph Bahn, Dean Bailey, Floyd Baker, Mrs. Glenn Baker, Mrs. B. Paul Baresch, Mrs. Emmett Bennett, Mr. and Mrs. George Blair, Mrs. Lucyle Bloom, Bob Bradeen, Ray Bradshaw, Mabel E. Brown, Art and Harry Buskala, Duane Carstenson, S. Cushman Clark, Earl Clarkson, Jack Coleman, Timothy Coleman and Art Colgan.

Art Davis, Mrs. Tom Davis, Dorothy Dippert, Bill Dodd, Mrs. Ed Dungey, Charles Edgerton, Mr. and Mrs. Fred Fisher, Joe Forest, Charles Foster, Henry Frawley, Sr., Colonel and Mrs. Charles Gardner, Tom George, Walter Gerbracht, George Glover, Elvie Gotfredson, Richard M. Grenville and Tom Guider. Mrs. Byron Hall, Kenneth Hanify, Edgar Harrington, Mrs. Chris Helmer, Lucille Hill, Mrs. Pete Hill, Don Hobart, Bernece Holly, John Honerkamp, Vernon Holst, Marie Huddleston and George Hunter.

Verla Jewitt, John R. Johnston, A. J. Kammerer, Albert Keffler, Mrs. Mabel Kiel, Margaret Kindler, Marshall E. Lambert, Norman Lambert, Vera Lambert, Ena Lancaster, Mr. and Mrs. Marc Lamphere, Marie Lawler and Waldo Lewis.

Vernon McCoid, Francis McDonald, Keith McDonald, George McDowell, Mr. and Mrs. J. F. McFarland, John McVey, Marguerite Melchert, Clyde Mitchell, Mr. and Mrs. John Moline, Art Moore, Ellen Mueller, Berneice Musekamp and Mr. and Mrs. Walter Mutchler.

Mrs. Meredith Nelson, Peggy Nelson, Mrs. James Newland, August Nonnast, Mr. and Mrs. Roy Norman, Charles A. O'Donnell, John O'Donnell, Ed Orestad, May Picore, Walter Plunkett, Ray Rathburn, Mrs. Alvin M. Reuppel, Mrs. W. A. Reynolds, Mr. and Mrs. Leslie Richards, Mr. and Mrs. Howard Riddell, Elsa Rowe and Dillon Ruland.

Mr. and Mrs. Walter Sandal, Bill Schekel, George Schieke, Dorothy Schnute, Mr. and Mrs. Ed Schriner, Linfred Schuttler, Phillis Seaman, Jim Secrest, Mr. and Mrs. Dennis Seiler, Mrs. Edwin Seiler, Stanton Selway, Avon Shaw, Mr. and Mrs. Leo Sherman, Mr. and Mrs. Allen Shoun, Lloyd Stewart, Mr. and Mrs. Calvin Stomprud, Mrs. S. T. Taylor, Layton Terrill, Claris Thompson, Ward Van Horn.

Joel Waterland, Mrs. Walter Weaver, Delores West, Kent Whittemore, Bill Williams, Emmett Wilson, Mary Ann Wilson and Mrs. Alfred Young.

Cover: Carolyn Cypher

## TABLE OF CONTENTS

# 1
# ALL ROADS LEAD TO DEADWOOD

It was 1877 and the Black Hills in Dakota Territory took center stage when they were officially opened to white settlement. Gold, that beautiful, shiny metal, known since antiquity, gave the cue to the nation that here in the gulches and streams were instant riches—the stuff of which dreams are made. The Hills were in the middle of settled country and in a comparatively short time pilgrims could make it in from any direction they chose—and in they came by the hundreds from Cheyenne, Sidney, Montana, Bismarck and Fort Pierre. A few lucky ones made their fortunes, although not always with placering, some stayed to make a living and others left, empty-handed, muttering that the new Eldorado was a big fraud. One of the lucky ones was a gentleman who took out six hundred thousand dollars worth of gold dust to Philadelphia and had it minted into twenty dollar gold pieces.

The nation learned of the gold deposits after Custer's Expedition of 1874 but a number of adventurers, explorers, traders and others knew of its existence long before, in fact several outfits were formed to go in the Hills before it was legally opened for settlement. As early as 1803 Volle, a French trader, had ventured "three hundred leagues up the Cheyenne River under the Black Mountains," and saw several Frenchmen already here, living with the Indians on Rapid Creek, the Belle Fourche, Redwater and Spearfish Creeks. And even before then the Spanish had been on the south fork of the Platte and perhaps they had ventured a bit further north and dug the prospect holes found by the earliest pioneers.

In 1823 James Bridger lived with the Sioux in the Black Hills and had seen "brass" in possession of the Indians . . . he later made several trips through the area as a guide. But Bridger had so many tales of the places where he had been and of the things he had seen—including stories of glass mountains and spouts of hot water in a place called Yellowstone that people didn't know whether to believe him or not.

The Indians picked up gold nuggets but they didn't realize their value at first. Father DeSmet, a missionary among them told the tribal members that white men highly prized gold and if they knew that it could be found in the Hills would come in by the thousands. He cautioned the red men not to reveal the secret to anyone. Father DeSmet had come to the Hills of Gold in 1840 starting from Chimney Rock on the Platte . . . the Sidney Trail of later years. He traveled west of the Hills; the following year he started from the same point and explored the country east of the Hills. He stopped in Hot Springs that was a resort of the Indians, at Battle Creek (Hermosa), and passed the site of Rapid City. He came close to Bear Butte, entered the Hills and proceeded up to Terry Peak and on to Devil's Tower. In 1851 he started out from Fort Laramie, then northwest of the Hills to Redwater, again to Terry Peak and back to Fort Laramie—with all these trips he had a good knowledge of the Hills.

In 1833 the Ezra Kind party was in the gold country, their story, "Got all the gold we could carry," scratched on the Thoen Stone found many years later near Lookout in Spearfish.

In 1853 and 1854 there was considerable mining done around Deadwood and some believe that in 1868 there was mining done through Choteau of the St. Louis Fur Company, certainly much evidence was found in later years—for instance in 1878 a sluice box was taken from Sawpit Gulch that was twenty feet below the surface. Near Bald Mountain was a caved-in tunnel with an ore dump indicating that the tunnel had been about thirty feet in length. A lead had been followed to the top of a hill with approximately fifty shafts. The area resembled a battle ground with numerous rifle pits and stone redoubts . . . perhaps a scene of a battle in 1852, the year estimated by a tree growing from the mouth of the tunnel.

There was also much evidence of a battle on claim Number 15 on Deadwood where an ax or hatchet of French style was found many feet below the surface. Numerous bones indicated that a fierce battle had raged here, likely between white explorers and natives. Near Crystal Mountain, to the south of Lead and Deadwood, was found a French type ax in later years.

Near Deadwood a pistol was picked up that had been manufactured in 1853, a chain was found in the Trojan area that had been deeply imbedded in a tree and throughout the area there were deserted cabins—many of them in a dilapidated condition in the 1870's. A number of trees had been blazed, mute testimony of early visitors. Some of the early arrivals may have been men who were a part of the 1849 California Stampede and made a one hundred mile detour to look over the Black Hills.

Many scientific and military expeditions were made to the area. In 1865 L. D. Waln came in with a regiment under Colonel Sam Walker—the party camped on Beaver, Iron and Sand Creeks, on the Belle Fourche and Redwater during their sojourn. In 1875 Waln came to Custer and that May drove the first bull train to Deadwood, or at least one of the first because several others vie for this honor. Early in the spring Samuel Tillotson brought in a train of merchandise to Deadwood coming in by the Custer Road. Citizens in the gulch, including Bill Gay, met Sam at the head of Two Bit Gulch and escorted him to camp with much rejoicing. This was Sam's first—and last trip—he killed the oxen and opened a butcher shop, selling the beef from fourteen to sixteen cents a pound. Another who brought in a large shipment to Deadwood in the very early days was C. V. Gardner.

In 1866-67 Martin Golden came to the Hills; he had been discharged from the army in 1865. He then came west with Colonel Irvin to establish military posts, roads, etc. He worked in Cheyenne and helped in establishing Fort Laramie. On his Hills trip he came in via Custer, Harney Peak, camped at the foot of Terry Peak and passed out through Whitewood Gulch. He said later that it was most difficult to get out and the party had to cut roads and build several bridges.

In the middle '70's the newcomers didn't care who had been here and when—they were interested in the present—and in the gold deposits. In March of 1876 there were fifty men in Deadwood, by April a townsite had been laid out and by fall the place was an established city. The first comers had come down the sides of the cliffs and mountains into Deadwood—one route was into Split Tail Gulch near the site of Montana City. It wasn't long before roads were established—most of them called roads by courtesy only. They were narrow, full of ruts and potholes, tree stumps and boulders and with a lack of bridges. Good roads or bad roads, freighting and transportation companies were established and even a few maverick outfits came in including one in 1876 when R. E. Thompson, with fourteen passengers, drove a stage from Salt Lake City to the gold camp. Mr. Thompson said he had offered to bring in the Hills mail from Fort Laramie but the postmaster there would not give it to him.

By 1877 Deadwood was an up-and-coming place, definitely the queen city of the Hills. There were several hotels including the Welch Hotel, Grand Central and the IXL. In August the Custer House opened a first class billiard hall, there were a number of stores, restaurants, a bank and a lot of homes. One of the biggest businesses was created by the powerful thirst of the inhabitants—there were nine breweries in the Hills that were happy to supply the brews to the 258 liquor dealers who also had large amounts shipped in from the states. The Eureka Bar had genuine imported champagne and the Red Bird Saloon was selling gold and silver watches—no doubt they had received them in payment of drinks they had sold to cashless customers.

The following year—1878—Deadwood was still growing. The Hazen House served two hundred dinners one Sunday, and other hotels and restaurants were also doing a landslide business. The merchants had many customers and one general merchant received 150,000 pounds of freight in one day and over one million pounds in a week. People in the camp were putting on metropolitan airs—a gentleman owned a parrot that was a non-stop talker, Ben Hazen claimed to have the only gold fish in town and one individual owned a Stradavarius "fiddle." Bootblacks and chimney sweeps were on the streets and a hand organ man was making sweet music—well, music anyway. Six kittens were auctioned off with two hundred people putting in a bid. One import wasn't popular—bedbugs—an eastern "luxury."

Deadwood supported approximately one hundred liquor establishments and in April one-half dozen more opened up for business, including the Blue Bird on Sherman Street. The Red Bird was going strong, the place next to the Cheyenne Stage Office was doing equally well. Milliard's Saloon and the Wide West were running faro banks and Madam Moustache was getting ready to open another one. There were a number of rude theatres and the Gem proprietor ordered a tent seventy by ninety feet with a portable floor that could be used for dances and variety theatre. Even with all the 'joints'—called 'hells' by the local newspaper, there was less ruffianism than expected and in May it was noted that no one had been shot the previous day or night—so things were getting rather "dull in the camp."

By the end of the year the gulch city had a population of five thousand—and five hundred and fifty of them were voters. The Deadwoodites observed, rather disdainfully, that Rapid City had only forty voters—the rest had gone to Rockerville that was a bonanza camp for a brief time. Deadwood had two churches, a school, two daily newspapers, a telegraph, two daily mails and excellent hotels and business houses, although one disenchanted person said he didn't think there was one thing pleasant or homelike in the place. He complained that the streets were narrow and dirty and the buildings huddled so closely that they couldn't get air. The yards were filthy and the opium smoking in Chinatown was a disgrace; however, this pilgrim did acknowledge that the scenery was just beautiful.

The last year of the '70's saw still more growth in the city and the six to eight brickyards in the vicinity were producing two million bricks for the building boom. The town matrons at their get-togethers observed that good servant girls were scarce, they were all a-flutter over two Paris dresses that had arrived and wished they could buy the silver tea set on display at Star and Bullocks. Imports no longer were only the basic supplies as they now included mahogany counters, soda fountains, sweet butter and wind vanes. The IXL was serving beefsteak or codfish balls with potatoes, hot rolls, tea and coffee—the steak for twenty-five cents and the cod fifteen cents. A roast dinner was twenty-five cents and corned beef and cabbage the same. Drinks were two for a quarter, a haircut thirty-five cents and a shampoo two bits. Oats brought eight cents a bushel, wheat thirty-five cents and corn ten cents. Butter was twenty-five cents and eggs fifty cents. Coal oil sold for a dollar a gallon, bacon fifteen cents and sugar six pounds for one dollar—people then would have flipped over most of today's prices.

The '80's brought more of the same for the gulch town—more people from the states, more merchandise for the businesses—and more businesses. Deadwood now had 656 voters and they weren't too happy with the growth of Rapid City that claimed 956 voters. However, explained the Deadwood people, the 956 voters included practically all the voters in Pennington County from a short distance from Lawrence County on one side and into farming country the other way.

The mining camp had very narrow—and dusty streets—but this wasn't new, although the residents did something about it now. On each side of Main Street was a water ditch and each business establishment had a water basin at the end of a long pole that they dipped full of water and sprinkled on the streets—not only to lay down the dust but also cool the air.

The leading hotels included the Wentworth, the Merchants and the Cosmopolitan. The best meals in town were fifty cents, board was from two to five dollars a week. The Deadwood Flouring Mill was producing flour under several brand names—the Board of Trade, Star, Diamond and Snowflake. The finer things in life weren't neglected—the opera house was often filled to capacity.

All in all Deadwood was an interesting city and proudly wore the title of a cosmopolitan place "with pigtailed Celestials and Ethiopian cousins thrice removed to artistic Italy's inartistic sons and unclassic wanderers from Greece—almost every nationality on earth."

Not only were there people from around the globe but many came from surrounding states—a large number came from Iowa including the Hough family who came to Dakota in 1882. Grace Hough, now Grace Stephens, was only a baby when they left Spencer, Iowa in April. Her father had traded his horse for two bulls and described them in an account of the trip. "A large roan who we called Lion and a dark red with a white spot in his face who we called Tiger. We made a frame for the wagon bows extending over the wheels made a cover and tent. Put everything inside . . ."

The party—it included several relatives—went only about four to five miles the first day to get "organized" for longer hauls. The next morning Uncle Romeyn (Mr. Hough's brother) gathered up his cattle and Mr. Hough his one white cow and the group set out again. The oldest children herded the cattle the entire distance. "We very soon encountered very bad sloughs which that country abounded with and which seemed to have no bottom," wrote Grace's father. "All the other wagons would get fast and I would have to unhitch and pull them all out although my load was much heavier than any of the others, but the bulls were very stout and very trusty. This was often repeated all the way to Sioux Falls until finally the bulls played out and we were in a dilemma what to do. The rest went on aways and left us. Romeyn came back and we concluded to lighten up the load. We got a big box and shipped it full to Chamberlain. We threw a heavy frying pan or griddle and some other things away. A pail of eggs was broken and wife was greatly provoked. We had a great time."

The party had a difficult time getting their cattle across the Big Sioux River as well as the Jim River near Mitchell. They went by the way of Plankinton and Kimball, both very small villages just started; however, Mitchell "was quite a small place." They camped at Chamberlain and Romeyn's hired man and another man were caught buying whiskey for an Indian, they were arrested, a "kind of a trial was held" and they were taken to

Yankton. Selling whiskey to an Indian in those days was against the law. The party planned to go straight to Rapid City but learned that the Brules were giving trouble and the road most difficult so they crossed the river on a flatboat and followed the west side up to Pierre. It was a terrible trip up and down bluffs—but they finally arrived in Pierre where they stopped for a couple days.

Mr. Hough wrote, "We had quite a time. Here Aunt Martha got mad at Romeyn and threatened to leave and go to Minnesota where her folks lived but Romeyn coaxed her to come on. We found it quite natural for people to get cranky after they have been on the road for a long time and there are but few exceptions." There were quite a few Indians in the vicinity but the party wasn't bothered. It was a chore getting over a number of creeks and at the Cheyenne the men had to wade up to their waists to get the teams across.

"Aunt Clara Marrs had a buckskin pony that was very balky and going up the long hill this side of the river commenced to back down on our team," wrote Grace's dad, and then continued, "We got out of the way and she went down for quite a way, but soon recovered, the road was very muddy but we kept moving until we got to the Shon (Shoun) Ranch. Here it commenced to rain. The ground was clear gumbo and covered with water we had to camp on the wet ground. We found here a big freighting outfit stuck fast in the gumbo not a spoke could be seen."

The party came up Elk Creek to Pleasant Valley and on up to Sturgis. Later Grace's family went to Brownsville where the father got a job at the Homestake Lumber Camp. Going up Boulder Canyon they looked so shabby that they weren't asked to pay the toll at the gate and at Deadwood a lady of the "fancy kind" gave one of the girls a pair of shoes.

The Houghs and many other early settlers came in with very little money and worldly goods but with hard work became substantial citizens in the community. Many of these pioneers built nice homes and farms for their families, they sent their children to college and they participated in community affairs . . . by their efforts the Black Hills was settled.

Of course, not everyone was an upstanding member of the community—there were still some of the Calamity Jane types around, notably Poker Alice who ran into a bit of trouble in Deadwood when she did some shoplifting. Lead people weren't too enchanted having her as a resident and she was asked to leave—she eventually settled in Sturgis. The Badlands in Deadwood's Half World was going strong and the unsavory Green Front and Gem Theatres were in the picture. Drinking had been curtailed in 1890 when statewide prohibition came in but package houses flourished.

The '90's ushered in the era of the railroads—the Burlington and Elkhorn appeared to join the Deadwood Central and Homestake's Black Hills and Missouri River Railroad. There was still considerable gold excitement with work being done at Two Bit, Bald Mountain, Yellow Creek, Garden City, Elk Mountain, Iron Hill and Polo Creek. There even was a stampede to Dotson—a camp located on the old freight road from Crook City. In east

Deadwood were chlorination works and smelters to process the ores.

Lead was growing with hundreds of people coming in, many of them immigrants from Europe, with a large number of them entering the employ of the Homestake that had become a giant. Rapid City wasn't standing still either and eventually became the largest city in western South Dakota. Sturgis and Spearfish were having growing pains as were other towns . . . and Deadwood lost its spot as the leading city in the Hills. But it is still a sturdy place very aware of its importance in the days of the gold excitement, with a history as rich as the treasures that were—and still are—in the "Black Mountains."

*Main Street, Deadwood in 1876, when it was "newly hatched." Soon it would have homes and businesses . . . plus saloons and theatres for entertainment. The Bella Union opened in September with a first class variety show; in July the Langrishe Troupe had a grand matinee one Sunday afternoon. In 1877 Al Swearingen's Gem opened in April with first class entertainment.—Photo courtesy Alumni Association, School of Mines.*

*Main Street of Deadwood when it was becoming a busy camp. In 1877 it was noted that the Sidney coach left at 6:30 a.m. and the Cheyenne coach at 7:00 a.m.—Photo courtesy School of Mines Alumni Association.*

*In 1878 Deadwood was the leading camp in the Hills. In June of that year there was considerable activity up Deadwood Gulch too—near Central City there were ten mills with a capacity of 310 stamps. The streets in the Northern Hills camps were in terrible shape—one observer said you could not tell the color of the horses pulling the Sidney coach—so covered were they with mud.—Photo courtesy School of Mines Alumni Association.*

*Merchants Hotel in Deadwood—one of the leading hostelries in the camp.—Photo courtesy School of Mines Alumni Association.*

*Freighting to Lead in 1878—the many mines and mills in the area gave a lot of business to the freighters.—J. E. Meddaugh, "Souvenir of Lead." 1892.*

*Evans Bull Train. At one time LaPlants Train of the Evans Line brought in 130,000 pounds of freight including a carload of bottled beer for Hermann and Treber.—Lawrence County Centennial Collection*

*A bull train—one of the many that brought in the merchandise and supplies to the Black Hills—everything from needles, pianos, wind vanes, sweet butter to giant pieces of mining equipment. —Black Hills Illustrated.*

*There was also overland transportation with power supplied by mules that were faster than bulls . . . however, they needed grain and hay as did horses while the bulls could manage with grass along the wayside.—Photo courtesy South Dakota State Historical Society.*

*Sometimes the streets would be blocked for hours when the wagon outfits came to the camp; some could not even leave their homes if they were too close to the streets.—Photo courtesy Alumni Association, School of Mines.*

*It wasn't easy to haul mining machinery up steep mountain trails—especially in wet weather when the roads were "bottomless." Here is a piece of equipment being hauled by a dray line.—Photo Black Hills Illustrated.*

***From placering to giant mills, gold and** silver were the key words in the development of the Black Hills. This scene is of a placer operation near Blacktail.—Photo courtesy Charles Martin.*

*The gold bullion had to be delivered from the mills—this is a bullion wagon no doubt on the way to a stage office. Wells Fargo delivered the bullion for a time; then the American Express. In later years it went on the Homestake train with Richard Bullock the guard.—Photo courtesy Phoebe Apperson Hearst Free Library; Meddaugh Collection.*

*Wagons of all types carried the freight—ores—grain. Studebaker was one of the leading manufacturers of wagons. A lot of gold left the gulches—in 1876 a party of miners left for the States with 2,500 pounds of gold ore.—Photo courtesy Phillis Seaman. Miles City Museum collection of vehicles.*

*The Deadwood-Spearfish coach owned by J. S. McClintock—the photo taken near the Pine Crest Park. The lady leaning out of the window is Mary Caretto . . . the little girl Allana MacArthur.—Photo courtesy Adams Memorial Museum.*

*The Sidney Stage office in Deadwood. The Northwestern was noted as having a most handsome office.—Photo courtesy Lawrence County Centennial Committee.*

*There were several stage lines from Deadwood to Spearfish; this was McClintock's Stage in front of J. S. McClintock's livery barn. In early days McClintock ran an omnibus between Central City and Deadwood.—Photo courtesy Adams Memorial Museum, Deadwood.*

*This was McClintock's Stage in Spearfish at the Spearfish Hotel. In early days many of the stage lines had four or six perfectly matched horses; McClintock had one white horse.—Photo courtesy Linfred Schuttler.*

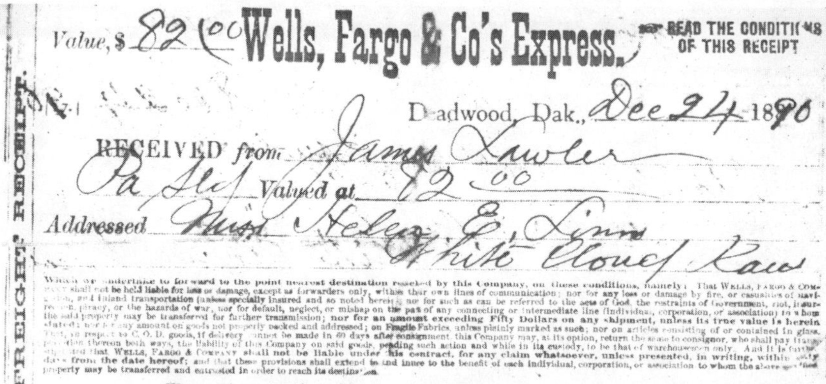

FREIGHT RECEIPT.

Value, $ 82.00 **Wells, Fargo & Co's Express.** READ THE CONDITIONS OF THIS RECEIPT

No. Deadwood, Dak., Dec 24 1890

RECEIVED from James Lawler

Pa Slip Valued at 82.00

Addressed Miss Helen E. Linn
White Cloud Kans

Which we undertake to forward to the point nearest destination reached by this Company, on these conditions, namely: That WELLS, FARGO & COMPANY shall not be held liable for loss or damage, except as forwarders only, within their own lines of communication; nor for any loss or damage by fire, or casualties of navigation, and inland transportation (unless specially insured and so noted herein); nor for such as can be referred to the acts of God, the restraints of Government, riot, insurrection, piracy, or the hazards of war, nor for default, neglect, or mishap on the part of any connecting or intermediate line (individual, corporation, or association) to whom the said property may be transferred for further transmission; nor for an amount exceeding Fifty Dollars on any shipment, unless its true value is herein stated; nor for any amount on goods not properly packed and addressed; on Fragile Fabrics, unless plainly marked as such; nor on articles consisting of or contained in glass. That, in respect to C. O. D. goods, if delivery cannot be made in 60 days after consignment, this Company may, at its option, return the same to consignor, who shall pay [illegible] thereon both ways, the liability of this Company on said goods, pending such action and while in its custody, to be that of warehousemen only. And it is further stipulated that WELLS, FARGO & COMPANY shall not be liable under this contract, for any claim whatsoever, unless presented, in writing, within [illegible] days from the date hereof; and that these provisions shall extend to and inure to the benefit of each individual, corporation, or association to whom the above [illegible] property may be transferred and entrusted in order to reach its destination.

The party accepting this Receipt thereby agrees to its conditions.

NOT NEGOTIABLE. For the Company, W. H. Wilson

Charges, $ Paid 80 c

*In 1890 James Lawler sent a package to his future wife in White Cloud, Kansas. It cost eighty cents to send the parcel valued at eighty-two dollars by Wells, Fargo and Company Express.—Courtesy Marie Lawler.*

*Calamity Jane taking a rest. She rode the stages, came by horseback or drove her own freight wagon on the trails to Deadwood.—Photo courtesy Alumni Association, School of Mines.*

*Deadwood Dick (Richard Clark) who lived in a cabin north of Deadwood; after his death he was buried nearby. There were many Deadwood Dicks—some say this one—the sixth one in the line—had been a shotgun messenger on the Ft. Pierre coach. And he could tell many exciting stories of his life—true or not. He lived in Whitewood in the house now occupied by Mrs. Walter Weaver. One time the newspaper commented that he had escaped many hair-lifting and diabolical plots and stage robbers . . . as depicted in early dime novels about a Deadwood Dick.—Photo courtesy Mrs. Peggy Nelson.*

*This tally-ho was widely used as a means of transportation—this one taking tourists to Sylvan Lake, a popular early-day resort.—Photograph: J. S. Sanford, Black Hills Souvenir, 1902.*

*A buckboard was a simple vehicle—a slat or boards placed between the front and rear axles. The advantage was its light weight—often used for carrying the mail. This picture was taken on the Main Street in Nisland.—Photo courtesy Claris Thompson*

*The train replaced the freight wagon. This photo was of the Northwestern that ran through Blacktail—taken near the present Athow house.—Photo Courtesy: Helen E. Cooper*

*The train was largely replaced by gas powered vehicles—trucks and cars. This is an early car—picture taken in Terry. —Photo courtesy Ed Johnson*

*The Black Hills was settled to the tune of rolling wheels—wheels of the stages, freight wagons, buggies, buckboards and tally-hos. Here is a wheel of one of those vehicles, patched and abandoned near Ten Mile. ". . . as a symbol of those days."*

# 2
# THE CHEYENNE TRAIL

In 1873 the country was hit by a financial panic caused by post-Civil War inflation and the outlook was bleak, especially in the eastern states where thousands of people were out of work when innumerable businesses had failed, factories were locked and banks had closed their doors. In the summer of 1874 when the nation was still in the grip of the depression George A. Custer and one thousand men—and one woman (Aunt Sally Campbell, a cook) were in the Black Hills of Dakota Territory on a military reconnaissance. In August men of the expedition discovered gold on French Creek near the present-day Custer. A messenger, Charlie Reynolds, was dispatched to give the news of the discovery to the world and he reached the telegraph office at Fort Laramie after traveling day and night over unfamiliar Indian country—and it is said with the shoes on his horse placed backwards to confuse the red men. The tidings were broadcast throughout the land—glad tidings too—a hope for the future.

Cheyenne, the City of the Plains, became a congregating point for would-be millionaires from New York City, Chicago, Salt Lake City, Denver, San Francisco, New Orleans and other cities, all anxiously waiting for word that the Black Hills had been opened for settlement. Not that all of them waited—large groups and individuals slipped in and were just as promptly moved out by the military and taken to Fort Laramie, Camp Robinson or Camp Sheridan where they were transferred to civilian authorities and in a short time released. One man reported that he had been moved out of the Hills four times in three months, returned each time and allowed he could keep it up as long as the army could keep it up.

A number of men were able to elude the military and remained in the gold country during the winter of 1875, including Robert Neal (Neil) who wintered on Sand Creek, there also were prospectors in Palmer Gulch and quite a few gold seekers in Deadwood Gulch. In September the Indian Peace Commission had tried to reach an agreement with the Indians but had thrown up their collective hands when the red men asked six million dollars for the Black Hills, plus "eighty years of boarding, clothing and washing for members of the tribe thereof." Finally in February of 1876 an agreement was reached and the Hills of Gold were legally opened for settlement—and the rush was on. At the height of the immigration it was estimated that every four minutes a "pilgrim" would set out from Cheyenne to make his fortune—hopefully—in the gulch country.

In the fall of 1875 a few brave freighters had risked the trip to bring in supplies. Some of them took the trail from Cheyenne to Fort Laramie, following Rawhide Creek, then to the headwaters of Old Woman's Creek, while others went from Fort Laramie to a point where the Cheyenne River crosses Wyoming into South Dakota. As did other trails, the Cheyenne Route followed early buffalo paths, the trail of the Indians, the early fur traders and of military expeditions. These routes were also followed by the stage that came in later.

The red warriors did not bother the large outfits too much but considered the smaller parties fair game, hence small outfits would congregate at Fort Laramie when coming in and at Custer City and Red Canyon when moving out in order to have the protection of large numbers. Several white men were killed and their stock run off . . . but not all the trouble was caused by Indians. Whitney, a freighter, was held up north of his camp near Old Woman's Fork and thought the robber had been Dunc Blackburn, an exceedingly busy road agent on the Cheyenne Trail. (This was in 1877.)

In January of 1876 J. C. Abbey's train came to Custer City with a full train carrying all sorts of merchandise, including an apothecary shop, blacksmith tools, crackers and bacon. Accompanying the train were two Chinamen from the Flowery Kingdom with supplies for "washee houses." A few other freighters came in during the winter months and by spring there were a number of trains. Gilmer, Salisbury and Patrick's fast freight train arrived in Deadwood City from Cheyenne in sixteen and one-half days, including a thirty-two hour layover at Hat Creek and eighteen hours in Custer . . . this was in June of 1876. In charge of the twelve wagons and forty-eight mules was C. A. Skinner, with a cargo that included sawmills, medicines, dental and surgical instruments, store goods and hardware. The next month B. F. Clark arrived with thirty-four wagons of general merchandise—and large outfits were needed to bring in the tons and tons of goods needed in the Hills. This included quartz mills—it was noted that there were five of them in Cheyenne waiting shipment. One mill came in during 1878 that weighed 450,000 pounds, with the freight charges totaling $27,000.00 and in 1879 there were three million pounds of boilers waiting for freighters to take them to the Hills.

Prices that had been sky-high, flour $20 per hundred and potatoes twenty-five cents a pound, fell down when the supply of goods increased.

By July flour was down to seven dollars per hundred and bacon was selling for seventeen cents a pound—and for the most part prices remained fairly consistent with those in the States. Sometimes when goods were in short supply the freighters coming in with these items would charge premium prices. In the winter of 1876-77 Johnny Spence did a bit of manipulating and cornered the kerosene supply and then charged $3.75 per gallon—a bit of wheeling and dealing. Even high prices didn't bring dismay to the prospectors who were getting twenty-five to one hundred and fifty dollars a day with their placering, although some sources believe this was a rather high estimate. Not every pilgrim hit a bonanza as in July of 1876 a lot of them were returning to Cheyenne, saying that the Hills were a big, fat fraud. An exception was one hardy individual who sold his boots for a grubstake and eventually left with $100,000 in gold bullion—and there were others who also made their modest and not-so-modest fortunes. It depended a lot on luck, the availability of capital, on pluck and perseverance and most of all, saying, "Come in," when opportunity knocked at the door. And with a minimum of government interference!

Most of the incoming freight was marked for Deadwood, the leading camp. In October of 1876 the gulch city boasted of two hundred business houses and three thousand people living one mile along Deadwood and Whitewood Creeks. The population included individuals from all walks of life, including a large number of the "rough element," that was responsible for the frequent shootings and killings. To keep the peace there was a police force that was kept mighty busy. Not everyone liked it quiet, including one drunken bullwhacker who jumped in the middle of Sherman Street, proclaiming, "Peace troubles my mind," and to create a little excitement threw a rock at another sinner. The two retreated post-haste when the law "hove into sight"—perhaps the freighter decided that peace wasn't all that bad—or maybe he didn't relish the thought of staying in the cooler until he sobered up.

There were other camps in the immediate area of Deadwood that also needed merchandise—including Gayville—a short distance up Deadwood Gulch—the oldest camp in the gulch. In the fall of 1876 they had 250 homes and thirty business houses including the usual saloons, a first class hash house run by John Smeltzer and a brewery that remained in business for many years. Other nearby camps were Anchor City, Golden Gate, Blacktail and South Bend. Further down from Deadwood were Fountain City, Elizabethtown and Montana City, up Whitewood was Pennington, the present Kirk. On Gold Run was Lead City that eventually became the largest city in the Black Hills—until Rapid City took over from a "hay camp" in the early days to the up-and-coming city it is today. Lead blossomed due to the discovery of the Homestake Mine by the Manuel Brothers and partners in the spring of 1876 and there were other mines and mills that promoted the growth of the Mile High City. The freighters were busy supplying all these camps with merchandise coming from near and far—in 1877 H. N. Orr's mule train came in with seventeen thousand pounds of goods that had come all the way from Massachusetts. Freight rates usually were from five to six cents per pound but they varied widely; in the summer and autumn they were lower and in the spring and winter higher due to the poor road and weather conditions. By 1879 it was noted that freighting was almost non-existent on the Cheyenne Route as the freighters preferred the shorter Bismarck and Fort Pierre roads. Two of the last large outfits on the Wyoming road were the Charles Clay outfit that changed to the Fort Pierre route in May of 1879 and the Kitcher and Hart train that had been engaged by D. T. Bramble for the season and then sent on the Fort Pierre road, too.

As early as December of 1875 the Wyoming Territorial Legislature had authorized the establishment of a stage line to the Black Hills. The Cheyenne businessmen thought it was a great scheme but they were reluctant to gamble money on it—there were just too many "coffee coolers"—as they called the Indians—floating around. George W. Homan, Jr. planned a stage line from Sidney to Fort Laramie and on to the Hills, then decided that Cheyenne would be a better departure point. His plans received the blessing of the legislature and Homan, greatly encouraged, purchased several stages and set up some stage stations. Agnes Wright Spring, in her book, "The Cheyenne and Black Hills Stage and Express Routes," states that a Red Cloud trader, J. W. Dear, set up several road ranches from Fort Laramie toward the Hills and thought he may have planned to lease or sell them to Homan. She listed the stations as Government Farm, Running Water, Old Woman's Fork of the Cheyenne, Horsehead Creek and the Cheyenne River, a day and one-half from Custer. Edwin A. Curley, in his book, "Guide to the Black Hills," said that Mr. Dear had located a "ranche" at Snake River for the accommodation of travelers, but his stock had been run off and one of his stage drivers killed so he had to abandon the stage line before he barely got it started. Snake River was about twenty-seven miles from the Platte, according to Curley. It could be that Dear had ideas of running his own line at one time but the Indian depredations made him change his mind.

While Homan was considering whether the Indians would leave his line alone, two men at Red Cloud—Frank Yates and Captain W. H. Brown started the first stage line to the Hills. They carried mail to Spotted Tail and Red Cloud Agencies and it was easy for them to extend the line to the Hills through Red Canyon. The initial trip from Cheyenne was made in January of 1876 but the press there gave the trip of February 3 a big bally-ho as the "maiden voyage," when J. W. "Doc" Howard, handling the ribbons set out from the Inter-Ocean Hotel with seven passengers.

There were several others who contemplated stage lines or pony express service to the Hills of Gold, including Joseph Mason who proposed a pony express with the trip to be made in forty-eight hours. W. D. Terry of Topeka, Kansas had ideas for a stage line—and this was mighty fine with the Cheyenne people who knew that competition would make for lower rates. The Hills people were pleased too—they wanted an up-to-date means of transportation at a reasonable cost and they looked forward to good mail service. Before regular mail service was established, receiving a letter was rather a haphazard thing for the gulch dwellers. Mail was entrusted to travelers coming this way and to the freighters—and without a doubt, much of it was lost in transit or if it reached the rightful owner was dirty, wrinkled and torn. It was good news when the Pioneer Pony Express Service was inaugurated from Fort Laramie in July by Charlie Utter and Professor O. A. Ingalls. But the Deadwoodites became suspicious when mail arrived that had been two to three weeks on the road—and yet they had to fork over twenty-five cents to receive it. An inquiry was instituted and it was learned that the postmaster at Fort Laramie would not give the mail to the pony express riders but would give it to a man named Clippenger who said he would deliver one-half of it—for a consideration—but refused to do so. Whenever a train carrying mail was about fifteen to twenty miles from the gulch city, a man would ride out furiously to meet it, take the mail sacks and then ride as the wind to Deadwood. When the people learned that the mail came only a fraction of the way by pony express they were highly put out although Ingalls and Utter said they had started the express in good faith and even stocked the line with good horses.

Then there was a change of postmasters at Fort Laramie and the new man, J. W. Ford, gave the mail to the Cheyenne-Black Hills Stage Company who carried it free-of-charge for a short time. In October the stage was coming in tri-weekly, in November semi-weekly and in March of 1877 when the stage company had the mail contract, daily service was provided. In January of 1880 the mail contract between Cheyenne and the gulch city was curtailed.

The Cheyenne-Black Hills Stage Company mentioned above was formed by John "Jack" Gilmer, Monroe Salisbury and Mathewson T. Patrick—all far-seeing, substantial builders of the frontier west. They sent H. E. "Stuttering" Brown to Cheyenne to look over the infant Yates-Brown Company and in February of 1876 they purchased the equipment of the fledgling outfit. Gilmer, Salisbury and Patrick (Matt Patrick later withdrew from the company—this was in 1878) became one of the most successful companies in the west with five thousand miles of stage and freight routes including lines in Montana, Oregon, Idaho and Washington. It has been estimated that they earned sixteen million dollars on their various routes in the years that they were in business.

Luke Voorhees, experienced frontiersman and stage line operator was instructed to lay out a stage route, hire drivers and establish stations. Agents were sent post-haste to St. Louis to buy six hundred stage horses and mules and thirty Concords were ordered from New Hampshire. In the interim surplus equipment that had been purchased from Wells, Fargo and Company, including $70,000 worth of coaches was put into use.

The route Luke selected was much the same as Highway 87 between Cheyenne and Chugwater. There it turned a fraction to the east to cross the North Platte at Fort Laramie where a government bridge had been constructed in 1872. Fort Laramie was one hundred and fifteen miles from the

Wyoming capital. From here the route went almost north to the present Lusk and on to the Hills almost the same as Highway 85 and the eastern branch of 85 alternate. Along the way commodious stations were built and freighters were contracted to handle consignments of supplies at the various stops.

Luke, Stuttering Brown as well as the others did a fine job and on April 3, 1876 three Concords rolled out of Cheyenne with Jack Gilmer of the company and one of the most experienced reinsmen in the west, handling the lines of the lead team. Each coach had eighteen to twenty passengers, some packed inside and the others seated on the top. The boots were full of baggage—these coaches could carry 4,000 pounds of express and average 8 1/3 miles per hour. The first regular coach reached Custer in May and then ran as roads allowed. In the spring the Indians harassed all the travelers, they raided the stock of the company and burned buildings, including three new buildings on the line. For a short time service was discontinued; however, by late June it was restored. Through the efforts of George B. Moulton the lines were extended to Deadwood and the rough, rutted trail from Custer was improved—in fact the stage company made new trails part of the way.

September 25, 1876 was a great day in Deadwood as on this day the regular coach arrived in town with nine passengers, including Alex Dunbar, C. V. Gardner, A. M. McMasters—and one lady, Mrs. R. B. Fay. The second coach had more members of the "fair sex"—Mary Griffin and Alice Boyd and one child. Passengers in both coaches said they had not seen one Indian on the way. The first coach had David Dickey as driver from Fort Laramie, who won the respect and warm admiration of all the passengers for his skill as well as for his gentlemanly behavior.

The trip to Cheyenne usually took four and one-half days and if the roads were in bad shape the trip could take up to six days. In 1878 when the route was in A-1 condition, the trip was made in forty-nine hours. Fare was twenty dollars for first class, fifteen for second and ten dollars for third—acceptable in gold dust at twenty dollars per ounce. In August of 1876 the stage office in Deadwood was moved to Judge Kuykendall's Store, in charge was Isaac Gray. Widely proclaimed was the fact that tickets could be purchased to any location in the United States east of Cheyenne via the Cheyenne-Black Hills Stage and the Union Pacific Railroad—civilization had come to the Hills.

Individuals coming in with high hopes of making a fortune or those leaving with "chins dragging on the ground," hit varied weather; one group coming in during the spring of 1878 had sunny skies because they sported "noses as red and shining as the headlights shining in a tunnel." In June of 1877 the stages and freighters were pelted with hail the size of a hen's eggs when twenty miles west of the Chugwater Crossing. The hail formed drifts five to six feet high that appeared as so many drifts of winter snow. In the spring and summer months the streams would become swollen after rainstorms and the stages and other vehicles would have to wait until the waters subsided unless they took a chance and swam over—a dangerous thing to do. At one time rain froze to the thickness of a man's arm and snapped the telegraph wires into a thousand pieces—the telegraph line was completed to Deadwood in December of 1876. Sometimes the prairies would be clear of snow but not the Hills and it would be necessary to change to sleighs to make the final lap of the run.

The famous and the infamous rode the Cheyenne Stage—if they could get a ticket. Scalpers would buy the stage seats and then sell them for ten dollars more to the bona fide customers. Booking passage—with or without the added "charge",—were actors and actresses, soiled doves and gamblers, businessmen and prospectors, professional men and children and women coming to join fathers and husbands. Single women came in in search of employment—or husbands. There were people of all nationalities including the Chinese previously mentioned and it was noted that Ski Hi, Sing Smal, Ko-Ket, Bi Dam and several other Chinese ladies and gentlemen arrived in Deadwood in June of 1877 in a large wagon drawn by four mules. At least the reporter was able to obtain most of the names because they usually had trouble with the Chinese spelling. Many times the names of the incoming whites would be listed individually and the Chinese lumped into ". . . and four Chinamen and two Chinawomen."

Most of the orientals set up laundries or ran small shops, a few joined the whites and did prospecting and placering and there were even a few Chinese cowboys. At one time the Homestake Mining Company hired a group of "pig-tails" to work on their Black Hills and Fort Pierre Railroad but the occidentals got their dandruff up and chased the Chinese back to town. The people from the Flowery Kingdom could work for minute wages and no white could subsist on the same pay. In fact it was said that the Chinese could manage to live on ten cents a day and some of them were able to accumulate modest fortunes. Doctor Pin Hoy, one of the earliest Chinamen in Deadwood—left for China via Prescott with considerable wealth that he had earned in the Hills.

In 1876 Jack Langrishe, well-known actor, his wife and twelve members of his troupe came to the gulch city. He built a make-shift theatre at the corner of Wall and Main Streets—this building had a canvas roof since shingles were hard to come by. In a short time he leased the Bella Union that was weatherproof and played to widely appreciative audiences with a repertoire that included "The Dumb Boy of the Pyrenees," "Streets of New York," that had seven acts, "The Dutch Lovers," "Flies in the Web," and other plays. Before Jack's own theatre was completed, he and his troupe performed in Central City, Crook City, and Rochford. The building wasn't in use very long when the Deadwood fire of 1879 destroyed it. Jack and Mrs. Langrishe moved to Wardner, Idaho; and in 1885 went to Murray where he worked two years on the newspaper, the "Sun," and was elected state senator.

Langrishe was noted as an excellent actor of the "old school," a poet and newspaperman—and a good-hearted soul always good for a touch from anyone who was down and out. When he moved it was a great loss in the entertainment world in Deadwood. There were a number of people who wondered how an actor of his caliber who certainly could have made it big in legitimate theater in the east would come out to the frontier. Perhaps he had a sense of adventure—or he may have had the idea of getting rich as he had a mining claim in the gulch that did produce well for a time.

There were others who came in, too, including James Irion in 1876. He was shortly after killed on False Bottom near the present-day St. Onge and is buried next to the St. Onge Rodeo grounds. Sergeant J. O'Leary arrived in March of 1878; he was the son of Mrs. O'Leary whose cow was said to have started the great Chicago fire. One exceedingly ugly girl started from Cheyenne but didn't quite make it; seems she had such big feet that she couldn't navigate the sharp corners and had to return to the City of the Plains. The Wyoming city was a nice place, the Deadwood newspaper hastened to say, even if it didn't have the best-looking girls, however, it did have docile young men—such gentle creatures. A man there was teaching them the Newport Glide but the others threw him out, so perhaps they weren't so docile after all.

A far from docile individual was Kitty Arnold who came to Deadwood with a wagon train—one of the first women to arrive in the mining camp. She wore a regular buckskin suit—complete with the pants, which was very unusual for a woman in those days. Kitty commanded universal respect even if she did go down the streets with her revolver, announcing, "I am as tough as any of you and don't you forget it." She may have earned the respect of the gulch citizens not only with her gun but due to the many acts of charity that she did—as did the famous Calamity Jane. Unlike Jane, Kitty became a respectable citizen—in 1892 she was living in Custer, had married and was rearing a family.

The fair sex was soon represented, not liberally, but they were here in early-day Deadwood. There were Calamity, Kitty, Fannie Garretson, Kitty LeRoy and the Valdi Sisters. The latter were trapeze artists and in August of 1877 a benefit was given for them at the Gem. At the Bella Union a line of beauties was doing the can-can. In the camp for a brief time was Red Jacket Woodward who had been born in New York State. She ran away from home with a gambler and the two traveled up and down the Mississippi for a number of years. In 1876 Red Jacket came to the Hills but soon left and spent her last days in Chadron. There also were many hard working women who ran restaurants and boarding houses, including Mrs. Hepburn who had a boarding house. Aunt Lou worked for the DeSmet people and then left to run her own boarding house in Sawpit Gulch. In later years she moved to Beulah. Mrs. Frenchy came to the gulch city in 1876 and did housework for the town's matrons.

While the newspapers were describing the newcomers and taking a few slings at Cheyenne, they were also full of stories of the gold shipments leaving the Hills. Captain S. M. West and son left for the southern stage terminus with a three hundred dollar retort of gold—this to show that there was gold dust in the area—as if anyone could doubt it after all the publicity in the nation's papers. Over eight thousand dollars worth of Hidden Treasure bullion was sent out, followed by an eleven thousand dollar shipment from other mines. Later in the month $31,000 was sent out—and each shipment got extensive coverage by the press. The eyes of the road agents popped reading these accounts and they got in the action and made the news, too. The coach was robbed three times on one trip and it was NEWS in June of 1877 when there were no robberies reported for two whole nights. The company found it necessary to place two mounted messengers to go with the coaches between Hat Creek and Jenney Stockade, a favorite stamping ground of the road agents.

In March of 1877 the stage was held up a short distance from the mouth of Gold Run Creek, east of Lead. There was spring in the air and a few patches of snow gleamed on the hill east of Whitewood Creek on that tragic night. The late evening was "bright as daylight," when Jim Berry, Joel Collins, Frank Towle, Sam Bass and Reddy McKemma stopped the coach with the usual command—"Halt!" Johnny Slaughter, the driver, had quite a time bringing the horses to a stop and had almost succeeded when they became frightened and jumped to one side. Reddy had a nervous trigger finger and shot Johnny and slightly wounded a passenger, Walter Iler, who was seated on the seat with the driver. Johnny fell to the ground and the horses took off. Iler was able to untangle the reins and brought the team to Deadwood.

The first men on the scene were W. Von Bodingen, Dutch Charlie Hornick and a man named Hanlon. Von Bodingen had left Lead after ten o'clock, went down Gold Run followed by his dog, Spotty. At the mouth of Gold Run he heard shots and in a few minutes the coach came swirling past. Von Bodingen stopped at a nearby cabin and asked the men there, Hornick and Hanlon, if they had heard the shots and seen the stage. All three went down the road, came across two blankets, two hats and Johnny breathing his last, his whip still clutched in his hand. They discerned five men climbing the hill on the east side of Whitewood. Shortly arriving on the scene was Seth Bullock, sheriff of Lawrence County, accompanied by several other men from Deadwood.

Von Bodingen stayed in Hornick and Hanlon's cabin until daylight and then went on his way. Later he learned that the passengers had at first thought he was one of the road agents and were prepared to shoot him until they saw Spotty—and the dog, without a doubt, saved his life.

Johnny was the son of the marshall of Cheyenne, a popular young man. His body was taken there and buried with one of the largest crowds of mourners ever seen in that frontier city.

People were understandably upset with the shooting and interested in apprehending the highwaymen. Frank Towle was arrested by Bullock, but released on the grounds of insufficient evidence. He later took part in a robbery at Old Woman's Fork and was shot by Boone May and J. H. Zimmers, messengers with the stage. Afterwards May heard there was a reward of

$200 on Towle "dead or alive" and he went back to the scene of the shooting, dug up the body, severed the head and presented it to the county commissioners for the offered reward who disallowed the claim.

Reddy McKemma went to Ohio where he bought a farm, was trailed there and placed in the local crowbar hotel. He broke out of jail with his guns blazing in February of 1878, but was re-arrested a year later and committed to the penitentiary in that state. Joel Collins, Sam Bass and Jim Berry robbed the Union Pacific in September of 1877 and this was their Watergate. All three were subsequently killed and their share of the money recovered.

Other highwaymen—and there were a large number of them—included Charley Ross and John H. Brown who stopped the coach four miles from Whoop-Up, a wild and wooly area in the early days. The driver was ordered to halt in the usual manner, the passengers told to alight in order to be searched. A man named Finn, a Union Pacific conductor, shot at Brown and the two desperadoes returned the fire. Pickings were slim—thirty-two dollars and two watches. The bandits came to Deadwood and Ross took the watch Brown had received in the division of the spoils and hid it "twenty yards from the corner of the white house on City Creek," the small stream that comes down from the Mount Roosevelt area.

Brown was mighty sick and a friend of his became worried that he might die and asked Lurline Monte Verdi to come and see him. Lurline had been married to an army surgeon and studied with him so was quite efficient at doctoring. She left her restaurant and went to see Brown, who still had the bullet in him. She extracted it and the man recovered his health. Eventually he was picked up by the law, tried and sentenced to twelve years in the penitentiary at Lincoln, Nebraska.

In June of 1877 the Cheyenne coach was stopped three miles beyond the Cheyenne River and forty miles this side of Hat Creek. While the seven passengers watched, eight men opened the Custer treasure box with a hatchet and then they worked on the Deadwood treasure box and finally blew it up with powder. After one hour they told the passengers they could continue on to Cheyenne and even gave them twenty dollars each to pay expenses. Two shotguns and pistols were returned after the ammunition was shot off. These highwaymen had a bit of Robinhood in their hearts but this type was certainly in the minority. It has been said that men were driven to horse stealing and robbing coaches because they couldn't find work, and it may have been true with some, but many of them had outlaw hearts and with ideas of getting rich the "easy way." And one man, John Babcock, was pressed into being a road agent by Fonce, "The Kid", Bryan and there may have been others who became men outside the law by association with those who had chosen to be robbers.

There was Joe Minuse—no doubt called Mexican Joe—who was alleged to be a spy for the highwaymen. He was shot by Boone May and Billy Sample in Deadwood when they were trying to get information from him and the mask slipped from Billy's face—and this sealed Joe's fate.

"Mex, you've seen too much," said Boone and reached for his gun as did Billy. In those days justice was swift, sometimes it miscarried and there certainly was no reading a crook's rights and the other legal niceties of today. Bang! and Joe was dispatched to plead his case before his Maker.

The stage was held up at Robber's Roost, east of Hat Creek, and later three of the gang were recognized in Deadwood. May wanted them arrested but Bullock was reluctant to do so. It ended that Webb, one of the robbers, had a running battle with May, jumped on his horse that was tied to the rack in front of the post office but before he had gone far the horse was shot from under him. The outlaw surrendered and later his partners were arrested at a corral. In time all three crooks turned over a new leaf and became exemplary citizens—Webb was a county official, G. W. Conners a trusted employee of Gilmer and Salisbury and C. P. Wisdom worked for a freighting outfit.

One of the busiest gangs was the Jack Wall and Dunc Blackburn outfit that held up the stages repeatedly, although in the summer of 1877 they were outwitted by a gentleman who hid several hundred dollars in an old shotgun that the robbers didn't even look at twice. Jack and Dunc often visited Crook City where they were accepted by citizens of the community. One time Seth Bullock was notified that members of the gang were in the village and he, a deputy and the informer had them almost "walled in" when Seth's gun went off accidentally, the robbers were warned and got away. Later they were captured by Scott Davis who tracked them down for three hundred miles; the two were convicted and sentenced to the penitentiary. In 1895 Wall had been released and was employed in an Omaha smelter and noted as a reliable and trustworthy employee. Scott Davis received the reward of four hundred dollars for the arrest and conviction of Wall and Blackburn.

The best known robbery was the Canyon Springs holdup in September of 1878, the stop not far from Jenney Stockade near the present Newcastle. The robbery took place in broad daylight which was unusual as the road agents usually preferred the darkness of night for their activities. A man named Red Cloud rode out from Deadwood to alert the gang that the treasure coach, the Monitor, was on the way. The stock tender, William Miner, was imprisoned in the grain room and the outlaws then leisurely took out the clinking between the logs. When the coach rolled in Hugh O. Campbell was shot and Galen Hill, a shotgun messenger was wounded. Although wounded, Hill was able to shoot one of the robbers, a Frank McBride who died later. Scott "Quick Shot" Davis, captain of the shotgun messengers, first fired from the coach and then ran behind a tree and shot from there. He remembered the treasure safe was supposedly impregnable for twenty-four hours and decided to go for help. He met Jesse Brown, Billy Sample and Boone, also messengers for the company, who wondered why the coach was late. When the group got back to the scene of the robbery they found the highwaymen had fled—with the contents of the treasure box that had taken them only two hours to open. Gone was the money, bullion and jewelry valued at twenty-seven thousand dollars.

Miner, the fellow who had been locked up in the grain room, had managed to get out and had gone to Deadwood to report the robbery and get help for Hill who had been taken to the Cold Springs Ranch. F. L. Babcock, M.D. went out at once and extracted three pieces of rib bone that the robber's bullet had driven in. Gale recuperated in Deadwood and by October was up and about and enjoying a cigar. The townspeople commented that he was such a slight-looking individual, frail and certainly bearing no resemblance to the famous robber-gobbler that he was.

An extensive campaign was instituted for the capture of the highwaymen . . . Carey, Doug Goodale, Al Spears, William Mansfield and "Big Nose" McLaughlin. The stage company offered $500.00 for the capture of each man and $25,000.00 for the recovery of the treasure. In the robber clean-up several men were captured who may or may not have been involved in the Canyon Springs holdup, including two men who decorated a tree seven miles from Jenney's Stockade and left there as a warning to others of their "ilk." Young Goodale was traced to Atlantic, Iowa, by William Ward, the road superintendent. Goodale's father was a bank president and just couldn't believe that his son was a road agent—but there in the window of the bank were two gold bars marked, "Homestake." Ward started back to the Hills with the prisoner who was accompanied by his father and an attorney and somewhere along the line Goodale jumped the train and was never found—and Ward lost his job with the company. Al Spears received a life sentence to the penitentiary at Lincoln, Nebraska, and Mansfield and Big Nose were hanged by a group of vigilantes when the two were being returned to Deadwood from Cheyenne to stand trial. Eventually most of the treasure from the robbery was recovered.

Tom Price, said to be the head of the gang, was not involved in the holdup although he was in the middle in other of the gang's escapades. Jesse Brown, Jim and Boone May and others captured him at Ten Mile and the outlaw was wounded. He was taken to a ranch where he recovered eventually and then taken to Cheyenne where he was booked in the "Hotel de Draper." He was tried and committed to the penitentiary at Lincoln.

Throughout the years the stage company took many steps to foil the road agents. They had a safe that they used as a treasure box, they got another safe that would take twenty-four hours to open and another that would take fifty-four hours although neither one lived up to its warranty. They fixed up two stages as mobile treasure boxes, designed with steel plates. In 1878 they changed the route to go through Horsehead because it afforded less hiding places for bandits. However, the greatest deterrent was the "stringing-up" of agents as already mentioned and the increase in the number of shotgun messengers. Then in 1880 the Sidney Stage started taking the bullion shipments, including the Homestake gold, and some robbers changed their base of operations; however, horse thieves, an exceptionally industrious group, remained in the Wyoming country. The stage company figured that most of the robberies were "inside jobs" and whenever a robbery occurred some of the personnel were sure to lose their jobs—guilty or not.

In 1880 the eastern mail came to the Hills via Fort Pierre—Cheyenne had lost the mail contract—but people in northwestern Wyoming still needed mail service, as well as passenger service. A semi-weekly buckboard went to Buckhorn near the South Dakota border west of Lead and the regular stage came to Deadwood via Horsehead.

At this time a new Eldorado appeared when gold and silver ores were discovered north of Fort Laramie to Running Water and the following year copper ore was found northwest of the fort. Scores of people left Cheyenne for the new mining country. Jeff McDermott was the first to locate in the copper country at Rawhide Buttes, followed by a company that spent $100,000 in building a smelter and opening a claim. Towns came into existence, including Brackettsville (Brackett City), McHenry City and Maysville or Mayesville.

In May of 1883 Gilmer and Salisbury sold their Cheyenne-Black Hills stage line to Russell Thorpe who moved the home station to Rawhide Buttes where he made a home for his wife and son. (At one time Thorpe considered moving to Deadwood and managing the Bullock Hotel.) No sooner had the property changed hands when the new proprietor ran into hard luck. That spring there was so much high water that many of the bridges on the stage road washed out. The Thorpe dwelling at Fort Laramie burned. The wet spring was followed by a hot and dry summer with fiery winds blowing across the prairies making it miserable for everyone. The hot summer was followed by a winter that produced one blizzard after another and heavy snow plagued the travelers. Thorpe coped with all these setbacks and the stage kept running. In February of 1884 the newspaper noted that the Cheyenne Stage left every day to Deadwood with lines to Gayville, Golden, Rapid City, Battle Creek and other camps . . . and did until December of 1885 when Thorpe established a line from Hat Creek to Rawhide to connect with the Chadron coach—the rails had crept to Chadron by that time.

As the railroad kept erasing the stage lines it had a series of camps at its head, some of which became permanent. Jenney's Stockade gave way to Newcastle and the stage station was moved to the newer city that at one time was noted as the wildest town in the west. Tubbstown flourished a bit to the east—but that is another story of a newer era.

The old stage coaches were sold, some abandoned on the prairies and others tucked here and there in sheds and barns. One of the treasure coaches, Old Ironsides—also known as the Monitor, was used by Buffalo Bill Cody in his Wild West show. He took it on his tours to London where it was admired by the Prince of Wales, in France it was looked over by that country's president, then it was on display in Germany and at the Vatican. Now it is resting at the Smithsonian Institute in Washington, D.C., a memorial to all the brave men who were instrumental in carrying the Black Hills gold to the railroad.

# 3
# ON THE TRAIL FROM CHEYENNE
## From the City of the Plains to the Gulch City

Cheyenne, born in Dakota Territory, sowed her oats in Wyoming Territorial days and became a sedate citizen when the territory became the state of Wyoming. It was in the territorial days when Deadwood beckoned on a golden horizon that Cheyenne was at her zenith as a sinful and lusty frontier town that gathered men to her bosom, only to send them on to the gold camps—of course, after outfitting them for the trip. Offered were miner's outfits, cooking utensils, provisions, mining tools, teams and wagons—a team of two horses cost from one hundred to three hundred and fifty dollars and a wagon to match cost an additional one hundred to one hundred and twenty-five dollars. In 1875 Cheyenne was the largest city in Wyoming, with four to five thousand people and indications of "solid comfort and growing wealth," much of it due to being the first major jumping-off point to the Hills of Gold.

Agnes Wright Spring in her "Cheyenne and Black Hills Stage and Express Routes," gives a good description of the various stops and routes that changed frequently. New roads were made to shorten distances, to avoid holdup artists, to give better service or to carry mail to new camps and towns. The stations were established as needed or new ones constructed and the old ones improved, the goal at all times being to give satisfaction to the patrons of the line. The Cheyenne stage and freight routes became famous—for carrying passengers, mail and freight but most of all the golden treasure out of the Hills.

The first stop nine miles from Cheyenne was the Nine Mile Stop or Dry Ravine that was opened in December of 1876. It was not a regular stage coach stop although it was patronized by freighters and other travelers. Here was a fine well with good water.

The first regular stage station was at Lodge Pole Creek, eighteen miles from Cheyenne where Fred W. Schwartze had settled in the early '70's and in the spring of 1876 constructed a hotel. The ranch was a favorite stop for many of the early travelers.

Twenty-eight miles from Cheyenne was the second stage stop at Horse Creek Ranch, usually referred to as Fagan's Ranche. Michael Fagan, the proprietor, furnished meals at fifty cents and could stable seventy head of stock. There were no trees at the place, only a house, saloon and barns.

Three miles up the road was Isaac Bard's road ranch that in 1877 became a regular station called the Little Bear.

Ten miles from Horse Creek and thirty-eight miles from Cheyenne was the Bear Springs Ranch with plenty of stable room, hay and grain—and the best of liquors sold anywhere. Good meals, too.

In the Chugwater Valley, fourteen miles further on were several ranches including John Phillip's Chugwater Ranch where good meals were served—he also had a well-stocked bar. Accommodations were available for fifty head of stock. Nearby was H. B. Kelly's Ranch and a regular stage station. Kelley had settled here in the early '70's and he and his Indian wife built up an excellent ranch, growing fine vegetables, good cattle—and trees. Kelly's brother-in-law ran the stage station for him. In 1884 Kelley sold the spread for one-half million dollars—a mighty handsome sum.

Past the Point of Rocks, ten miles from the Chugwater, was Chimney Rock—here was a post office. Then Hunton's Ranch, Bordeaux, was passed that became a relay station in June of 1876. Hunton gradually acquired fourteen thousand acres of government land—which he fenced, every last acre.

Four miles to the northeast and about eight miles from Chimney Rock was Chug Springs used as a stopping place in the early days—the ranch built up by a man named Patton. It is said that Johnny Owen won it in a poker game . . . a man who in his lifetime killed twenty men in self-defense. Johnny was a slender, good-looking man who dressed exceedingly well—in fact his hobbies were reputed to be good clothes and good horses.

Seven miles north Eagle's Nest was passed, the nearby trail used as a cutoff in the Oregon Trail days. The stage station was about a mile from the mouth of Eagle's Nest Canyon. Next point was the Six Mile Ranch, a relay station and a place of numerous killings. It was located six miles south of Fort Laramie. Three miles from Fort Laramie was the Three Mile Station, a meal stop for the stages, excellent meals offered for fifty cents. In earlier days there were eight two-room cottages here with "ladies of the evening" occupying them, including one Calamity Jane. The place had several owners.

Fort Laramie was the dividing point of the south and north divisions of the Cheyenne Trail. The fort, first known as Fort William, then Fort John and finally as Fort Laramie, was on the north side of the river where the Laramie River entered the North Platte. The place had been established as a trading post in 1834 and purchased by the government in 1849. It had neat buildings mainly of adobe and stone. The headquarters of the Cheyenne-Black Hills Stage Company were in the Rustic Hotel. In 1890 the fort was noted as a veritable deserted village.

The government bridge was two miles from the fort, over it went the traffic to the Hills. It had been constructed in 1875 by the army at a cost of $10,500 . . . and is still in existence. In fact it is the oldest bridge in Wyoming and believed to be the oldest military bridge west of the Mississippi River.

The next stop, not a regular one, was at Ten Mile or Swing Station. Fifteen miles from the fort was an important mile-post, the Government Farm, where experiments had been carried on for raising vegetables and grain for the army but abandoned before 1872. A stage station was established here and in 1879 there was a house, stock, a good spring and two hundred acres of fenced land.

Rawhide Butte was a towering mark as the trail entered the Jay Em section of Wyoming. A mile and one quarter from the butte was the Rawhide Buttes Station situated along a creek—a favorite camping place in the early days. In 1876 J. W. Dear had built a home here that was soon burned by the Indians; in August of 1877 a stage station was erected with Mr. and Mrs. Amasa Lowery in charge—Mrs. Lowery was a sister of Luke Voorhees. Amasa and Sarah came to this country in 1881 from Michigan and homesteaded land near Jay Em—the town is named for his brand. He sold the place in 1892 to a man named Johnson who in turn sold it in 1898 to the Harrises and descendants of that family still run the place. Amasa was the postmaster at the station until 1895.

Leaving the Rawhide Butte Ranch was the road to Rawhide Springs, a good place for the early freighters. Next was a station at Running Water (Niobrara). Nearby was a rock formation called the Cardinal's Chair.

A station was built and operated by Jack Madden, the station about a mile west of the present Lusk. In 1879 Indians killed two freighters nearby although one was able to get to the ranch before he died. His partner was mutilated but not scalped. The Indians ran off with the freighter's four mules and two horses.

The earliest Cheyenne Trail went to Red Cloud—both from Fort Laramie and from Rawhide Buttes. In September of 1878 the stage went to Horsehead from here when the newspaper reported that there would be less hiding places for road agents. The road also went to Horsehead from the next station—this was the Hat Creek Station, one of the best along the line with a good bridge, sufficient grass and plenty of good water. It had been constructed by John Bowman and had a hotel, post office, brewery, bakery, butcher shop and a blacksmith shop. In July of 1878 the stage passengers commented that a huge herd of buffalo here was headed Deadwood-way; these animals hadn't been seen in the vicinity for fifteen years. A later coach brought passengers who reported that the shaggy beasts had disbanded.

At Hat Creek the Black Hills trails divided—the Jenney Stockade

Route to the north and the Indian Creek route to the east and on to Custer. Mrs. Spring listed the Harding Ranch, fifteen miles from Hat Creek and Indian or Prairie Springs on this route. Hyman Palais in "A Study of the Trails to the Black Hills Gold Fields," South Dakota Historical Collections, traces the route from Hat Creek as winding through a number of water courses, under the shadow of a great overhanging bluff of yellow earth for about eight miles to Indian Creek. The trail followed the creek for fifteen more miles going gradually toward higher ground. Seven miles from Indian Creek was a site known as Robber's Roost—a happy hunting ground for Indians and stage robbers—no wonder pilgrims would sigh with relief whenever they traversed this area without incident.

There was a station located on a high bluff overlooking Indian Creek in Nebraska, according to Palais, but a sliver from the southwest corner of South Dakota . . . about twenty-five miles from the Indian Creek Station. From here the trail went to Igloo, then to the site of Edgemont—known in earlier days as Dudley—and in this area the tracks can easily be followed.

Freighters in the early days spent their fifteenth night from Cheyenne at Hat Creek, sixteenth at Indian Creek, seventeenth at lower Indian Creek, eighteenth at the Cheyenne River.

Igloo was the site of a government ordance depot during the war years, then closed and employees transferred. When it closed it hurt nearby Edgemont financially—but today, once again, the town is booming thanks to Security Industries, uranium northeast of there and repairing coal cars. The coal is mined in Wyoming and sent to Arkansas—handsome coal cars go through in large numbers. The population of the town is 2,300 people; they expect a population of 7,000 in a short time. In the old days the problems were Indian attacks, horse thieves and road agents—today it is schools, water and streets to accommodate the increased population.

Northeast of the town was Camp Collier or the Red Canyon Station, a fort established to keep the white men out of the hills and later to protect them. In 1972 the site of the fort could easily be seen, a mound that apparently was a base for cannon and the usual glass and metal bits that are usually found at the old stations. The place was abandoned in June of 1877 when the route went north to the Jenney Stockade. The newspaper reported that in January of 1877 two cutoffs were made between Red Canyon and Fort Laramie—the first at Indian Creek that saved miles of travel and the other south of Bowman's Ranch on Hat Creek that made a saving of six miles.

Red Canyon a few miles further made an admirable gateway to the Hills but it was also a most dangerous place with Indians and robbers having many hiding places among the rocks and boulders. There is one place where the bandits had filed a crack in a big rock so they could spot their potential victims without being seen themselves—this rock was blasted over fifteen years ago during the uranium boom. In a narrow part of the canyon the Metz party was ambushed; Metz had been a baker in Custer and then decided to go back to Cheyenne. The whole party was killed—Mr. and Mrs.

Metz, the driver of the team and a black servant who wasn't found immediately. She was found with an arrow in her back and five dollars in her hand. It is said that the men who found her buried her—and with the five bucks bought beer . . . perhaps as a wake?

There were many others who met an untimely end in the picturesque corridor. It is a beautiful area with a creek meandering between the red walls and here and there supporting a stand of cottonwoods. In later years people settled in the canyon and then moved on again; remaining are several empty cabins; however, there are several ranchers still living in the area.

About twenty miles from Custer was a station known as Spring-on-the-Hill. In later years a concrete reservoir has been constructed to store water from the spring. In the spring of 1876 a party of six men was ambushed there and J. D. Hunter was killed. A man named Coyer was wounded and soon died.

Spring-on-the-Right was the next stop south of the later Eighteen Mile Ranch. The station was in Pleasant Valley, a first-class ranch with good stabling and plenty of hay. A bar offered the "choicest brands of liquors, cigars and tobacco." Proprietors were Mishand and Raynard. Pleasant Valley is a beautiful place with much more rain than Edgemont and Red Canyon to the south. It also has a lot of interesting history, for here was the home of Tom Richardson, the brother-in-law of Doc Middleton. There are several of the original buildings still standing on the ranch but the original home has been torn down and a new house is at the site. No doubt, Doc visited the place at various times. In the valley is the Heumphreus Ranch (the "h" is silent) that was a stage stop in later days when it was known as the Twelve Mile Ranch. Toward Custer-way and toward Four Mile is the site where the last three buffalo in the area were killed. Four Mile was also known as Moss City.

The next stop was Custer and for some passengers their final destination. In April of 1877 the stage arrived in the city in the evening, the passengers ate supper there and then set out again arriving in Montana City by daylight. The route changed frequently as already mentioned—in the early days it went to Hill City, Golden City (Sheridan), Camp Crook (Pactola), crossed Box Elder, Elk Creek and Two Bit and followed Whitewood up toward Deadwood, crossing Smith's Gulch, Split Tail, past Montana City, through Elizabethtown, Fountain City and Deadwood.

The station near Hill City was about three and one-half miles south of the camp, located near the present Rafter J Bar (earlier known as the Walker place), and that now belongs to Tom George. The original buildings are about one-quarter of a mile from the bar where there is a neat white house that embraces the original log cabin. There is a picturesque barn that has a small room on one side that evidentally was the grain room and to the south is a modern addition. On the north side of the barn is an overhang where the hay may have been stored or perhaps as a protection for horses. Between the house and the barn and a bit to the south is a shed that was a cafe at one time. The old stage road came here from the southwest and then went on to Golden City.

Golden City was soon the possessor of a new name—Sheridan—that in 1880 was noted as a lively place but people were looking for a short life for the flume on Spring Creek. And a short life for the paychecks of the prospectors, thanks to a couple of faro games going on in camp. A boarding house was in business with seven boarders making up the "clientele." Sheridan now lies under a lake and on some days when the water is clear and the day is bright, you can see some of the old buildings at the bottom of the lake. The lake was built in 1939 by the Civil Conservation Corps.

The old stage road ran fairly parallel with the present road . . . and near the North Landing at Pactola Lake you can still see the old tracks. Pactola was first known as Camp Crook and J. C. Sherman had the stage station. It was a two-story building painted a greenish-gray. He had a store here and when travelers stopped overnight they may have slept in the barn—in those days most everyone carried their own bedding; the barn was a big frame building. The station was up for many years—kept up by Mr. Sherman and sons Charlie and Frank. The town of Pactola now lies under Dike I of Pactola Lake—including the old stage station.

On Middle Box Elder was the stage stop operated by Mr. and Mrs. Gilbert Towers who came here from Colorado in 1876. The charge for overnight guests was $1.25 for the night—the men slept on the floor and the women in the bedrooms on the ground floor as well as in bedrooms on the second floor. In later years a new house was constructed on the east side of the old building but Mr. Towers didn't like it and stayed in the old building that he much preferred. Besides the traveling public stopping here and enjoying Mrs. Towers' good cooking, there were other distinguished visitors including Seth Bullock and his friend, Theodore Roosevelt. The Ernest Rantapaa's now own the ranch. From here the road went toward the present Brownsville, down the "Long Hill", into Two-Bit and to Whitewood and up that creek to Deadwood.

In the spring of 1877 the route from Custer was changed—the road laid out by William Ward and others. It came to the Twelve Mile Ranch, Spring Creek, Log Cabin House (Gillette's), Mountain City (Deerfield), Reynold's Ranch, Elkhorn, Bulldog, Ten Mile, Pennington and Deadwood.

Twelve Mile today is the site of the Bradeen Home . . . and instead of being twelve miles from Custer it is nine miles. The original stage coach hotel was a four-room log house with two rooms downstairs and two upstairs . . . the building constructed of hand-hewn logs. We don't know who operated the station but in later years Charley Vandelier homesteaded here and at one time school was held in the building. Water was from a well in the front of the house and there was a deep spring a bit to the east and for good measure water was available from Vandelier Creek that went through the place. The Bradeens moved here in 1928 and in the '50's tore down the log house and the logs were as solid and sturdy as when the building had been erected.

There are patches of prairie scattered throughout the Hills; one of them is Gillette's Prairie, named for Uri Gillette who came to the area in the early days. He constructed a roadhouse on the stage road—a substantial two-story building with many rooms—each numbered. There was a well-stocked bar for the convenience of guests—and excellent meals were served. The building, of hewn logs, without a doubt would still be standing if it hadn't burned down a number of years ago.

The white-bearded Gillette was a most ambitious man, he not only ran the road ranch but also operated a dairy business and sold the butter at one dollar per pound in Lead. He did a lot of hunting and butchering, selling the meat in the area. After Gillette sold his ranch, he retired down south, later he entered the State Soldier's Home at Hot Springs where he expired. The early-day pioneer is interred in Hill City Cemetery.

A good history of the Gillette Prairie region has been written by Margaret Kindler, who lives with her husband, Leo, three miles north of the old stage station.

The stage road went to Mountain City, the present Deerfield. A good stopping place here in the '70's was Miller's Ranch that is now at the bottom of Deerfield Lake, another of the picturesque man-made lakes in the Black Hills. At Miller's was a stockade for protection from the Indians in 1876—at one corner of the enclosure was a grave with the bodies of three prospectors killed by the red men. A nearby ranch was Garvey's Ranch mentioned in the early papers when coarse gold was discovered on Gold Run . . . which triggered a gold rush. Even earlier Custer had come this way.

The next stop was on Reynold's Prairie where Joe Reynolds established a ranch. Joe had come to the Hills in 1875 from Kentucky by horse-back—but was removed from the area by the army. Later he established mining claims on the prairie and built the ranch where Mrs. W. A. Reynolds lives today. He had a toll road that went a bit west of the present road and came out less than a mile from the Melchert Ranch on the Black Fox Road. On the Nahant road is a marker of the Cheyenne Trail. At Reynolds the original barn is still up—remodeled of course but the other buildings are of more recent vintage.

The Elkhorn Stage Station, called Twenty Mile Station in the early days was near the early camp of Montana City and the later Nahant where McLaughlin had his Tie and Timber operations. The station was up the hill from the present Wolff home—first was the barn built against the bank and higher up on the other side of the road was the stop, a big white house, that burned many years ago. Mrs. Albert Melchert remembered that the barn had stalls with the horses' names at the head of each stall—Jack, Bill, Nell, Bert, etc. When she and her husband lived at Elkhorn in the '30's they remodeled the barn and then it burned in 1932.

The Bulldog Ranch apparently was not a regular stop for the Cheyenne Stage—it was too near the Ten Mile and Twenty Mile Stations. In the early days Madame Bulldog had the ranch—it was on the opposite side of the road of the present ranch occupied by Timothy Coleman. Madame was either German or Bohemian and kept a place that was rather unsavory.

George Glover's father did freighting in the early days and remembered his dad telling him that one time when his outfit was camped nearby he warned his men not to go to the saloon as they were leaving early in the morning and he didn't want them to get in any trouble. During the night there were shots from the saloon but Glover never learned what had transpired there because he and his men took off early, early in the morning. Later Madame had a saloon in Deadwood and in the '80's a newspaper item noted that "Madam Bulldog, lately of Deadwood" was dealing stud horse poker in Miles City.

In 1880 Timothy Coleman's father remodeled the home and made it a comfortable place to stop. The previous spring he had fenced and broken a portion of the valley at the head of Little Rapid and raised a good crop of oats. In 1882 the newspaper commented that Mrs. Coleman knew how to put on a good feed—at one time the Bulldog Ranch was a stop on the Custer-Deadwood Stage Road constructed and maintained by Daniel Rathburn. There also was considerable traffic to the west toward Newcastle. One of the original buildings of the Bulldog Ranch, said to be saloon, has been moved across the road and is southwest of the present home.

Ten Mile was constructed by Daniel Rathburn who had helped build the stage road from Cheyenne to Deadwood. The house and barn were owned by the stage company and in later years Rathburn became the owner. In 1877-78 he constructed the toll road from Custer to Deadwood as already mentioned, but the venture proved to be unprofitable. The Ten Mile Station was south of the later store and post office and there was said to be a watering tank for the passersby below the present Leslie Richards home. The earliest road went down Whitewood Creek, to Pennington and Deadwood; a later road went to the northwest, to the ridge and down Summit Hill to Whitetail Creek, Central City and Deadwood . . . there also was a road from Whitetail to Pennington.

Ten Mile became known as Englewood and in later days became an important railroad station for the Burlington. In 1892 it had 1,500 people when they were asking for a post office to be called Aken.

Pennington was just east of Yellow Creek where it joins Whitewood and has a history going back to 1876 when prospectors were checking out the streams for gold dust. By the following June there were sixty people in the camp and five hundred in the surrounding area and the place soon sprouted stores, eating places, a saloon, four quartz mills and several saw mills in the vicinity . . . plus the Sutro and Savage Tunnels on the steep hillside. The owners of the tunnels were searching for the Homestake lead and missed by a minute fraction.

The newspaper told about a new road from Custer; instead of the old route the new one would go more to the west through Cold Springs, Twenty Mile, Ten Mile and Deadwood.

The Buskala Brothers today live on a ranch west of Bulldog—when they were youngsters they could follow the stage trail as the trees were waist-high at that time. Members of the Harvey family, who lived near the Buskalas, also remember the old trail from Cold Springs, that went eleven miles to the Besant Place. It approached the ranch from the southwest, went through the South Meadow, circled to the west to the station and then left by the North Meadow.

The stage inn at the Besant Place was a long, low building, part log and part frame. Samuel Besant ran the station and later built a new house next to the inn. The gentleman arrived in the Hills in the early days from England and acquired three to four hundred acres of land until the Forest Service cut him down to one hundred and sixty acres. He raised grain and hay and had cattle—some believe he crossed his cattle with buffalo and used the hides for covering the furniture that he made. He was an excellent carpenter and built everything on the place himself—the furniture, window and door frames with the windows and doors to fit the frames. The barn was an interesting building with an extremely steep pitch on the roof, estimated to be at least fifty percent . . . which was excellent as this is heavy snow country. The highest parts of the barn at the ends were thirty-two feet high, the barn built of wide, wide boards with four inch battens. Water was from a well.

Mrs. Elsie Wood, a granddaughter of Mr. Besant, remembers visiting the place as a young girl . . . the old inn was still up. In 1929 Gladys and Ed Dungey moved here and all the buildings were still up; however, very little is left at the site now—the foundation of the house, some stone work, a cellar opening and pieces of broken glass, tins and crockery. In the house was a trapdoor to a cellar where the valuables of the travelers were placed for safe-keeping.

The trail left the Besant Station in a northerly line, Smith's Peak comes to view, the trail went on a ridge past Nigger Jim's Gulch, Keogh Draw and Cousin Jack Draw—the latter leading to the east and Keogh to the west. A short distance from the rim north of the Buskala Ranch was a fork in the road, with one fork going to the west toward Ward's Draw and the other to the east, to the later Dumont, Ten Mile and Deadwood.

An early newspaper item mentioned THE FORKS eleven miles out of Deadwood and one mile from Englewood. During a bad storm some men went out to meet the Cheyenne Stage but went only as far as the Forks as they didn't know which road the stage had taken. The routes may have changed "officially" but the stage drivers often took the road that was easiest to travel according to the time of the year and if the "old road" looked better they would use it. The freighters, of course, didn't have to follow scheduled routes and if they saw a new shortcut trail and if it proved to be a good one, others would soon use it, too.

In June of 1877 the northern route from Hat Creek was opened. The trail went to Old Woman's Fork where Sourdough Dick was the stock tender.

The next station was May's Ranch on Lance Creek—Jim May was a brother of D. Boone May. Here the trail went slightly northeast, crossed the Cheyenne River and went to Jenney's Stockade, fifty-six miles from Deadwood. It first was under the proprietorship of James Mathewson who sold it in November to a Mrs. Scott. It was a breakfast and supper stop and advertised that it had comfortable beds for the passengers and good stabling for the stock. Starting in April of 1877 the stages started night runs to Hat Creek from here with mounted messengers on the coaches to foil the road agents that swarmed in the vicinity.

North of the Jenney Stockade was a beautiful area with green meadows. At Beaver Station or Beaver Stockade Station (the present LAK Ranch) "Frenchy" was the stock tender. In about ten miles the Canyon Springs relay station was reached—here was a big barn with quarters for the stock tender. Two miles down the gulch was the Cold Springs Ranch, that was on the junction of the trail from Custer and the trail from Jenney Stockade. Between Cold Springs and Deadwood was a stop listed as the Milk Ranch.

In the autumn of 1877 a new route, touted by the newspaper as being much shorter, was opened that crossed Spearfish Creek nine miles north of Cold Springs, went down Icebox Canyon, past the Valley Station in charge of Henry Fosha and the Cheyenne Crossing of the Spearfish. Today at the crossing is a store where groceries are sold and gasoline is pumped . . . the original buildings were across the road. From here the road went to the Little Meadow Divide, Ten Mile, Whitetail, Poorman's Gulch, Central, Gayville and Deadwood. In April of 1878 there were heavy rains that made washouts on the Central Road and the people had to be transported from Central City to Deadwood by horse—and didn't arrive in the gulch city until 10:30 a.m. In July the coach went back to the Pennington route.

In March of 1878 the Cheyenne Route was listed as Ten Mile, Twenty Mile, Jenney Stockade and Hat Creek . . . this schedule given in the Black Hills Daily Times. In October the mail was changed to go from Rawhide Buttes to Horsehead and this closed the Hat Creek Station. This road to Horsehead necessitated a new stop—the Bluff Station thirty miles west of Red Cloud. In 1880 the mail from Rawhide went to Buckhorn twice a week—after all the people to the north of Rawhide needed mail delivery. It was changed to weekly service in 1881. Buckhorn is just over the line in Wyoming from South Dakota and has been moved several times—it was moved to its present location on the modern highway about five years ago. The barroom is part of the old post office, according to Marie Huddleston who runs the Buckhorn Saloon.

In March of 1882 Luke Voorhees who had been running the stage between Cheyenne and Hat Creek bid for the mail contract between Hat Creek and Spearfish via Hot Springs and Custer and Deadwood. In a few days he received notification that he had received the contract and planned to run a line of four-horse Concords from Cheyenne.

Shortly there was notice of another mail contract from Hat Creek to Spearfish that went via Jenney Stockade, to the northwest toward Inyan Kara, to Sundance and then to Spearfish in the Black Hills. A similar road had been discussed as early as May 1878 when a road was planned from Inyan Kara, six miles below the Stockade, Sundance Mountain and Spearfish

City and they were pretty sure of getting the stage off the old road. They didn't of course, but years later this route finally came into use by the company.

In 1883 Gilmer, Salisbury and Company sold the stage line to Russell Thorpe and in February of 1884 the new owner announced that coaches left every day running through Custer, Battle Creek, Rapid, Golden, Gayville and all mining camps. Changes were coming in fast, made so by the railroad companies that were laying rails closer and closer to the Hills. In 1885 a line was established from Rawhide and another from Hat Creek to Fort Robinson to connect with the Chadron Stage.

The route of the Cheyenne stage and mail routes changed as conditions demanded—new towns and camps, new gold discoveries, a shorter route or a better one—a tangled skein of byways and throughways with the old trails ever becoming fainter.

In May of 1879 the Cheyenne and Black Hills Stage Company announced full service to the Hills with the shortest wagon roads, trails or bridle paths reaching all camps either by Cheyenne or Sidney coaches or their branches. The mileages as follows were from Deadwood.

Beaver, east—20 miles
Bear Gulch, west—40 miles
Belle Fourche River, northeast—30 miles
Central City, southwest—2½ miles
Centennial, northeast—5 miles
Castleton, south—30 miles
Custer, south—56 miles
Caves & Falls of Whitewood—5 miles
Crook City, northeast—10 miles
Crow Peak, northwest—22 miles
Cold Springs, southwest—32 miles
False Bottom, northwest—7 miles
Forest City, northwest—43 miles
Galena, east—12 miles
Germania, east—25 miles
Harney, southeast—61 miles
Hayward, southeast—64 miles
Hill City—54 miles
Inyan Kara, southwest—60 miles
Jenney's Stockade, southwest—56 miles
Lead City, southwest—4 miles
Mountain City, south—30 miles
Oil City, southwest—67 miles
Potato Gulch, northwest—42 miles
Pactola, southeast—35 miles
Rockville, southeast—59 miles (Rockerville)
Sand Creek, northwest—45 miles

Spearfish—12 miles
Spearfish Falls and Canyon, northwest—14 miles
Sheridan, southeast—42 miles
Sitting Bull, south—40 miles

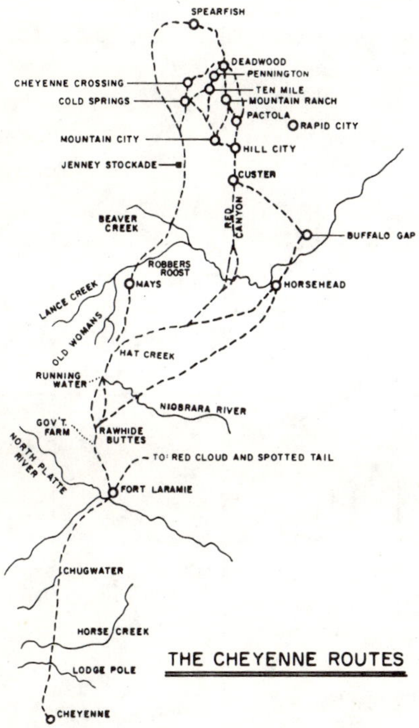


THE CHEYENNE ROUTES

THE CHEYENNE ROUTE
as given by Edwin Curley

# eating stations going to Dwd.
* eating stations going to Chey.

| | Miles | Remarks |
|---|---|---|
| Cheyenne | | |
| 9 Mile | 9 | Good Water |
| Pole Cr * | 9 | " |
| Horse Cr | 10 | " |
| Bear Spgs. | 10 | " |
| Chugwater # * | 14 | " |
| Huntons # | 14 | " |
| Eagles Nest * | 12 | No water |
| Wells Station | 14 | |
| Ft Laramie* | 3 | |
| Brackenridge # * | 4 | Good water |
| Govt farm# | 12 | " |
| Rawhide Butes* | 12 | Alkaline |
| Run. Water | 14 | Good water |
| Hat Cr. * # | 14 | Good Water |
| Indian Cr# | 16 | Good Water |
| Alum Springs | 14 | " |
| Red Canon* # | 15 | " |
| Sp.on Hill | 12 | " |
| Pleasant V. | 14 | " |
| Custer City # * | 9 | |
| 12 Mile | 12 | |
| Mt. City # | 12 | |
| Rapid Cr. * | 12 | |
| Hd.of Wtwd. | 12 | |
| Deadwood | 3 | |

Total:292 Miles

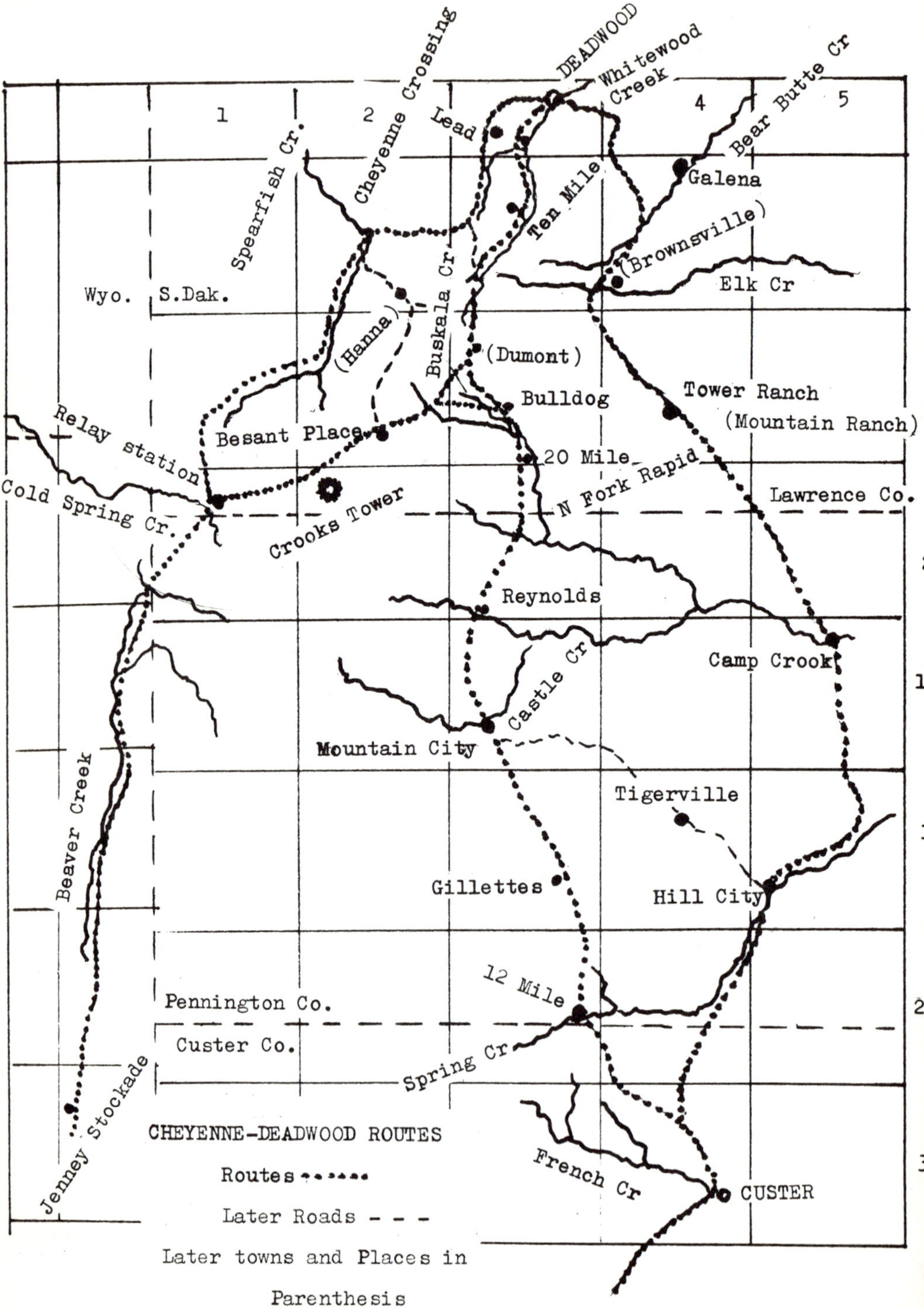
DEADWOOD
Whitewood Creek
Cheyenne Crossing
Lead
Bear Butte Cr
1
2
4
5
Spearfish Cr.
Ten Mile
Galena
(Brownsville)
Elk Cr
Wyo.
S.Dak.
(Hanna)
Buskala Cr
(Dumont)
Bulldog
Tower Ranch
(Mountain Ranch)
Relay station
Besant Place
20 Mile
N Fork Rapid
Cold Spring Cr.
Crooks Tower
Lawrence Co.
Reynolds
Camp Crook
Castle Cr
Mountain City
Beaver Creek
Tigerville
Gillettes
Hill City
12 Mile
Pennington Co.
Custer Co.
Spring Cr
Jenney Stockade
French Cr
CUSTER
CHEYENNE-DEADWOOD ROUTES
Routes
Later Roads
Later towns and Places in Parenthesis

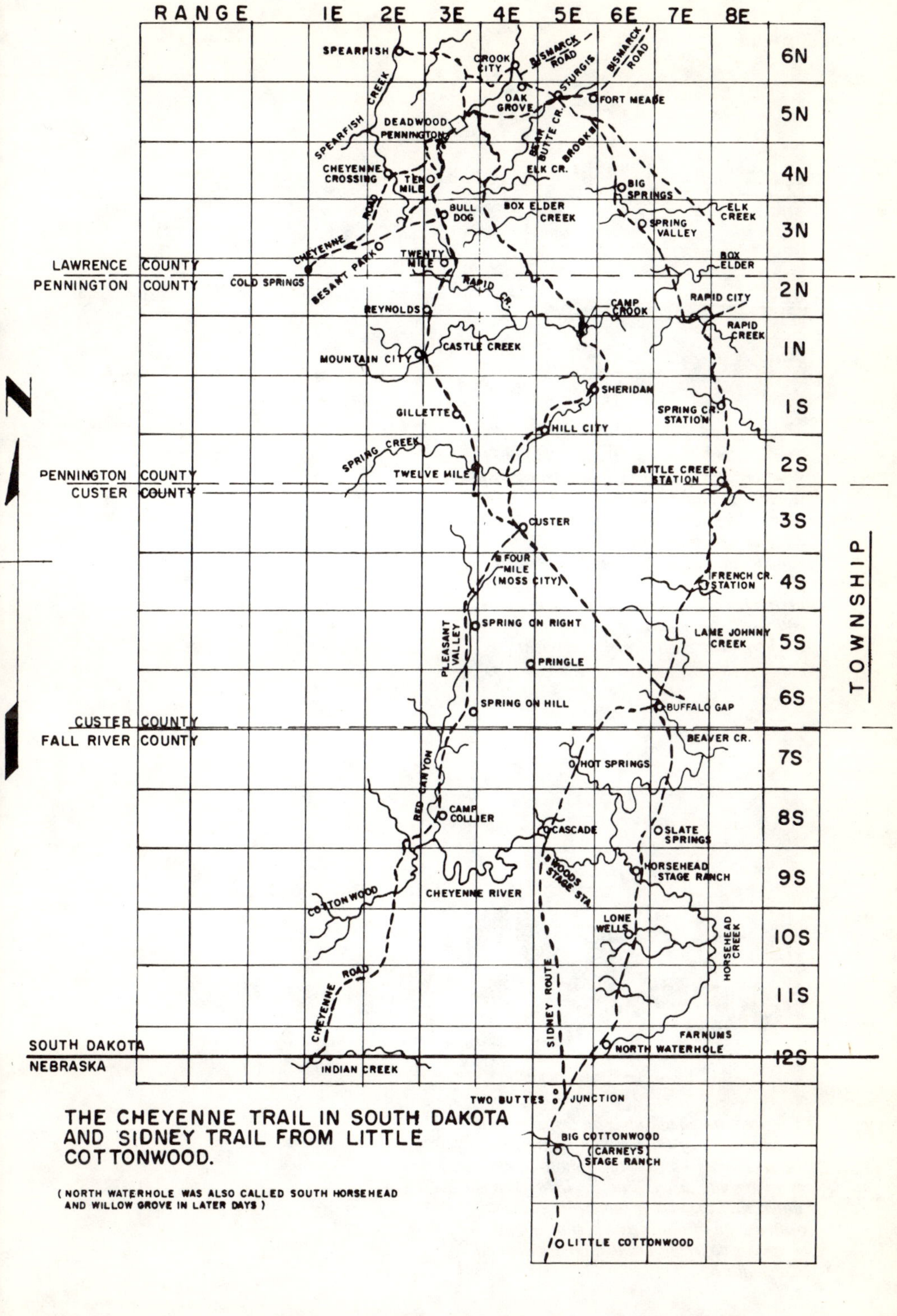


THE CHEYENNE TRAIL IN SOUTH DAKOTA AND SIDNEY TRAIL FROM LITTLE COTTONWOOD.

( NORTH WATERHOLE WAS ALSO CALLED SOUTH HORSEHEAD AND WILLOW GROVE IN LATER DAYS )

*The original Rawhide Buttes Station. It was partially burned by the Indians, then re-built. In January of 1877 the Cheyenne Stage Company was building a large number of ranches offering excellent accommodations, stock was being distributed and inside of two weeks they expected the coach to be running from both ends.—Photo courtesy Wyoming State Archives and Historical Department.*

*The treasure coach used at Douglas in a re-enactment of the Canyon Springs robbery. Horses—as in this photograph—were usually used, but in 1878 it was noted that the coach came to Deadwood with a span of mules as wheeling animals. —Photo courtesy Deadwood Centennial Collection.*

*The Cheyenne Stage came through the present site of Edgemont—this sign is at the edge of the town showing the path of the old trail.*

*An abandoned cabin in Red Canyon. The canyon was a gateway from the Cheyenne River area to Pleasant Valley and Custer. It has a lot of Indian markings on the canyon walls. In later days robbers liked to hold up travelers here—the odds were on their side with all the hiding places available.*

*Site of the Tom Richardson Ranch. Tom was a brother-in-law of Doc Middleton, horse-thief Number One.*

*The Heumphreus Ranch that was a stage stop in later days . . . the place is owned today by Mr. and Mrs. Ed Schriner who are planning a new home on the site.*

*The placque on the large rock tells us that the last three buffalo in Pleasant Valley were killed nearby.*

*The Gilbert Towers home on Middle Box Elder that was used as a stage station by the Cheyenne Stage Company when their route went past here. In time a new house was built next to this building but Mr. Towers much preferred the old building. The ranch is now owned by the Rantapaas. Photo courtesy Mary Bickle.*

*From Pleasant Valley the stage went to Custer where a stop was made at the elegant Occidental Hotel in 1877—then it went on to Hill City. A stop was made at a station where the Rafter J Bar is now located. The original house was of logs now covered with boards . . . this photo is of the station barn at Hill City.*

*The original Uri Gillette Ranch on Gillette Prairie that was used as a stage stop by the stage line. The house burned in later years.*

*This is the house today at the site of the Gillette Ranch.*

*The Reynold's Ranch on Reynolds Prairie on the way to Deerfield. Joe Reynolds came to the Black Hills before it was legally opened for settlement and the army escorted him out . . . but he soon was back.—Photo courtesy Mrs. W. A. Reynolds*

*The original barn at the Reynold's Ranch . . . it has been kept in repair or it would have tumbled down many years ago. Near here was Reynold's Toll Road.*

*The road from Reynolds went toward Twenty Mile . . . a sign faithfully records that this was part of the Cheyenne Trail.*

*The Twenty Mile Ranch site near the early town of Montana City and the later town of Nahant. Going down the hill the house was near the road on the right hand side, the barn a bit further on the left and built near the bank.*

*This building is said to be the saloon from the Bulldog Ranch—moved from the opposite side of the road and now is next to the creek on Timothy Coleman's Ranch. In the late '80's silver was found about four miles from here on the old Davey Salt Road near the head of Little Rapid.*

*The original Jenney Stockade building that has been moved to Newcastle and serves as an information booth for the Chamber of Commerce.*

*Buckhorn, west of Lead. In 1880 a semi-weekly buckboard brought mail here from the south, service was increased to three times a week. In 1882 the newspaper reported that the stage was taken off the Sidney route and once again was coming on the old route.*

*The Cheyenne Stage came through Icebox Canyon in later days and passed Hells Gate Canyon where there was a road to the Tinton Country—one of three gateways.—Photo courtesy Charles D. Ricker*

*This road from the south is on a ridge, said to be part of the Cheyenne Trail . . . in later years it was used as a cattle trail. This photo taken west of the Buskala Ranch . . . Art Buskala is the gentleman in the photograph. When there was a lot of traffic from the Hills to Newcastle much of it went west of the Bulldog Ranch and past the Buskalas and Harvey Ranch.*

*The Ten Mile Ranch was the later day Englewood. The road came from the southwest from the Bulldog Ranch, through the town. There was a watering trough below the Richards and Peterson homes, said to be used by the early travelers. This building was the stage stop.*

*The original Besant Stage Station site—Mr. Besant on the horse. This building was built much like modern-day hotels with many outside doors.—Photo courtesy Mrs. Elsie Wood*

# 4
# THE SIDNEY TRAIL

In the middle 1870's a Sidney visitor noted the Nebraska city as a "fast" town. In and around Spring Street, he said, were large encampments of greasers, Germans, H'Ingliher, tropical Injun, left-handed grog eaters, handsome Scotchmen and Celts (he evidently claimed them in his ancestry), and many of that detestable race known as Yankees. He went on to say that "one meets them at every turn, of a different hue and they tote a definite and not always fragrant breath."

Many of the bearers of the "not always fragrant breath" and others with a more quiet aroma were ready to take off to the new Eldorado in the Black Hills of Dakota. Some set out on horseback, others with wagon teams and still others walked with freight trains, paying a small sum to the freighters for carrying their belongings. Passengers rode with the fast freight teams, were charged nine dollars and furnished their own grub and did their own cooking. Later they rode the stages. The time it took to reach Deadwood varied widely, depending on the condition of the roads, the weather, the route selected, the Indians or the "knights of the road."

The Sidney the pilgrim left behind had come into existence in 1867 when the Union Pacific laid glittering rails here on its way to the west. At first the place was a sleepy little hamlet with a few saloons, gambling places, a round house, a couple of stores and a handful of homes, but then in 1874 it became a freight depot for the Indian agencies, Red Cloud and Spotted Tail, and it started booming. After gold was discovered in the Black Hills the boom continued—in fact accelerated when it became a jumping-off place for the freighters and travelers looking for a good route to the Hills. A lot of publicity was given the town by Joseph B. Gossage of the Sidney Telegraph who listed the advantages of Sidney and the Sidney route. Columbus, Grand Island, Kearney, Plum Creek and North Platte put in strong bids but never made the grade with the traveling public.

The biggest stumbling block on this route was the Platte River that was almost impassable from January to June and then when the water level fell off in the summer and autumn, the shifting sands were most treacherous. There was a ferry here but it wasn't too satisfactory and many of the travelers preferred going by Fort Laramie where there was a government bridge. Then to the pilgrims rescue came H. T. Clark of Bellevue and associates who built a beautiful bridge across the Platte near the present Bridgeport . . . forty miles from Sidney. The bridge was completed in December of 1876, a sturdy sixty-one truss over two thousand feet long and resting on deep-driven piles. There were excellent approaches at either end. On the south side were a store, hotel, a post office and a corral and on the north end was a two-story blockhouse. The place, called Fort Clark, was frequently occupied by soldiers to guard against the Indians and white brigands. A toll was charged of two dollars for a two-horse team and driver and three dollars for a four-horse team and driver.

The Union Pacific people also gave the Sidney route publicity by distributing pamphlets far and wide extolling the glories of the gold camps and the advantages of the Sidney route . . . which didn't set too well with the Cheyenne boosters. They protested by saying that they, too, were on the Union Pacific and would appreciate some publicity for their route. However, the railroad promoters were realists and knew that travelers wanted a shorter road to the Hills—and Sidney was fifty miles shorter.

A multitude of people answered the siren call and arrived in the Nebraska City to take off for the Hills of Gold. They were the same type of people as Cheyenne had attracted—miners, prospectors, mill men, merchants, doctors, freighters—many of the freighting men settled in the gold camps after a trip or two. There were Chinese, female sinners, actors, actresses, lawyers. Deadwood welcomed most of them with open arms, especially the show people. The local press gave them considerable publicity and even noted that Madam Biscuit with the Gem Variety Troupe dislocated her garter and lost her poodle "Hand," when the Sidney coach upset this side of Crook City . . . this was in 1880. Reminds one of the Hollywood stars being written up in magazines today; their children, what they eat, their pets . . .

The Gem and the Green Front were early day "joints" in the Badlands of Deadwood . . . the city's "half-world." They had legitimate performers and raised money for worthy causes as in 1884 the Gem had a ball to get money for the Irish Relief Society. The "upper crust" in the gulch city considered them disreputable as they also had gambling and the second floor was a "girlie" house, in fact one of the proprietors caused a major scandal when it was learned that he enticed girls into white slavery.

Early in 1876 J. W. Dear, a trader at Red Cloud, had attempted to establish a stage line to the Hills, and had constructed a "ranche" at Snake River and elsewhere; however, the Indians ran off his stock and one of his stage drivers was killed. Dear had to discontinue the line before he barely got it started.

The brightest star in the transportation history from Nebraska was H. T. Clark, the bridge builder. He was a mover, a rustler . . . and that didn't mean a stock thief. In those days a rustler was a man of great energy and enterprise and Clark fitted this to a "T". The United States government could not send mail to the Hills as it was still Indian country and any mail sent from one army post to another was the business of the military. Clark directed dispatch riders and government personnel carriers between Sidney and Red Cloud, Fort Clark and Fort Robinson and it was easy for him to send mail and express to Custer and later to Deadwood. The first mail was carried by pony express riders—including H. G. Rockfellow and Charles Nolin who was killed by the Indians a short distance from the present Sturgis. Later mail was carried by stage that also carried passengers and express. A stage line wasn't necessarily stocked with coaches—frequently the vehicles were buckboards, a tally-ho, buggies and wagons—the latter often on runners when there was snow on the ground. Clark did have coaches—later on anyway.

Special envelopes were offered for sale that cost ten cents plus the regular stamp. These envelopes stated: "In consideration of the ten cents for this envelope and of which payment its possession bears evidence, the undersigned agrees to carry from the Union Pacific Railroad at Sidney, Nebraska to Custer City and Deadwood, Dakota Territory (and such places as his route may supply) or from above places to the Union Pacific Railroad at Sidney without additional charge. H. T. Clark."

The Black Hills Daily Times informs us that Clark's Sidney Centennial Express left Deadwood on the first return trip to the Nebraska city on July 22, 1876, with the idea of completing the maiden voyage in ten days. It made good time and Clark decided to run a weekly express, leaving Deadwood every Tuesday and arriving in Sidney on Saturday. It paid to advertise, then as now, and an early ad proclaimed, "The Sidney Short Route to the Black Hills is 237 miles to Deadwood. Ranch, water and wood accommodations good along the line. The Sidney and Black Hills United States mail coaches travel over this route, carrying Clark's Centennial Express."

A news item of October 1876 reported that the Sidney coach with four through passengers arrived four days out of Sidney, the quickest time ever made from the Union Pacific. The paper also complimented Messrs. Clark and Company for showing a degree of enterprise worthy of all good words and works. In December the Sidney and Black Hills stage was running three times a week between Sidney and Red Cloud and once a week from Red Cloud to Deadwood. Ranches were being built or improved at Water Hole, twenty-five miles from Red Cloud, another at Horsehead, a distance of forty-five miles from the agency and one at the entrance to Buffalo Gap, sixteen miles from Horsehead. The fourth ranch was at Battle Creek (the modern Hermosa), twenty miles from the Gap. The stage time had been cut down to sixty hours from Sidney to Deadwood and it was hoped to cut it down to forty-eight hours driving day and night. In August the fares had been set from Chicago to Deadwood at $49.25 first class, $38.00 second class and $28.00 third class.

January of 1877 was a mild month and the freighters were able to keep their wagons on the road much longer than normally—the first week of the month twenty-seven freight trains came in. The stages were bulging with passengers—on three different days forty-one people came in one day, thirty-eight another and a third day forty-six. They found Deadwood a bustling town lording over a number of smaller towns and camps in the area. In the spring there was an increase in mining activities and a number of new discoveries made including forty eight claims at Nevada Gulch . . . and a stampede developed when a large group of people went there. A new and enterprising city was growing between Gayville and Golden Gate where buildings were going up—this was the budding Central City, that soon boasted of a hotel and four stores. A new town named Cleveland, one mile up Whitewood, was having speculation in town lots and a number of buildings were going up.

Mines were sprouting all over—the Pecacho, Caledonia, American Flag, Aurora, Fairview, Ophir, Great Western, Hidden Treasure, Alpha and Silver Chief—to name only a few. Mills were constructed to process the ores—the Cunningham, McLachlin and Castel, the Pinney where a clean-up of over four hundred ounces from one week's run from the Wolsey Mine was made. There were the Potter and Powers and the Pearson that had the first Troy scales in the area, placed in front of the mill and capable of weighing six tons.

Deadwood was losing its frontier image and getting to be a settled place—no longer was the shotgun and revolver the arbitrator as it had been up to this time—now the law was represented by local police officers. Not that there wasn't a vestige of frontier life—there was a shooting at Gayville but the people didn't know what to do with the gunman as there wasn't a jail there nor one in Deadwood. This was soon remedied when a freighting outfit brought one in.

In September of 1876 James "Modoc" Stephenson was in Deadwood regarding a stage line from Sidney that would connect the places in four days time. The line, the Western Stage Line, was formed in partnership with Captain W. W. Marsh, who in December of 1876 brought in the first stage from the Nebraska city with five passengers billed to Deadwood. Returning to Red Cloud he found that the company had no stock there and he was obliged to buy new animals to complete the journey. In an interview—many years later—the Captain said that at the time he thought it was a hard way to serve the Lord and was still of the same opinion.

In January of 1877 Mr. Stephenson was in Deadwood to make arrangements for the stage to come in tri-weekly. It wasn't done evidently as in March the captain came to the gulch city regarding the same thing. Shortly after Gilmer, Salisbury and Company, the big wheels of the Cheyenne stage business purchased the interest of Modoc Stephenson. They had observed the growing importance of the younger line and thought it would be a good move to get a piece of the Sidney transportation pie. They purchased the equipment in March and by April the line was running regularly.

Gilmer, Salisbury and Company were awarded the mail contract and had it for many years . . . when they had it from April to September of 1877 they were paid $17,000 to deliver the mail six days a week. In 1880 Congress failed to make an appropriation for mail delivery and the stage company delivered it only once a week. Finally Washington woke up, made an appropriation and the mail continued coming in . . . weather permitting. Bad weather could hold up the mail for days and when it finally would come there would be bushels and bushels of it. One time twenty-two sacks of mail piled up and another time 2500 pounds . . . but usually the company provided good service. The Crook City people didn't agree with this—they would get their Deadwood newspaper only two or three days a week by the Sidney stage and had some highly uncomplimentary things to say to M. Willsie, the agent for the line.

Traveling by stage—or private conveyance—was never a milk run, bad weather made it worse many-fold. In 1878 Mr. and Mrs. L. B. Bearey and family of four children almost drowned when crossing the Cheyenne at Snyder's Ferry. This was in April and melting spring snows had widened the river to a giant one hundred and fifty feet. The ferry capsized and the father and three older children were carried downstream to a bar and Mrs. Bearey with the eight-month old baby was held up by her hair by the ferry boat operator until they too floated to the bar. There were campers nearby who brought them to their camp where they dried off and no doubt soothed their shattered nerves. And also gave thanks to the good Lord for sparing them—it had been a mighty close call.

Smaller creeks would become raging rivers—and even dry gullies would fill to the banks during storms. In 1881 a swollen Snake Creek delayed the coach and Judge Barron's trunk was reported to have "drowned." In the wintertime animals were often trailed over the roads to pack down the snow and make them passable for vehicular traffic. Rains and snows would create mudholes and there was much griping about the state of the roads—and Deadwood got its share of criticism . . . especially the Chinatown road where it was so narrow that only one coach could pass between the houses and Whitewood Creek. One time a Celestial full of enthusiasm and rice whiskey threw a pack of fire crackers into the street and the leader of the stage horses whirled around and the coach jackknifed. Fortunately a gentleman who had been nearby grabbed the bits of the horses and saved the coach—and the passengers—from a swim in the creek.

The muddy streets in Deadwood and on the roads was something fierce. In 1878 the freight and express teams would be stationed on the street crossings in order to have more solid footing—some swore the streets were bottomless. One time when the Sidney coach came in an observant gentleman said one could not tell what color the horses were—so covered were they with mud.

However, after a hard trip through mud, snow or sleet, the passengers arriving on the Sidney Stage could go next door to the stage office and have a few drinks at Newton and Wallace's Saloon. If they didn't want anything alcoholic they could stop at the Oyster Bay and have cocoa, tea or milk, or a bowl of oyster stew. The early Hills people were inordinately fond of this seafood and one time when there was a strike on the east coast and no oysters available, it hit the gulch dwellers hard—in the stomach.

Bad weather did not stop the immigrants—this was accomplished by the growing popularity of the Bismarck and Fort Pierre lines. On top of the lessened passenger load the company suffered several calamities . . . the first in 1878 when the stage barn at Sidney burned. Lost were five horses and a number of coaches with the loss amounting to four to five thousand dollars. In 1879 the barn in Deadwood burned—this was in the big fire that left a large portion of the city in ashes. The stock was moved into Doctor Pease's Livery Stable on Upper Main and the office into the Werthheimer Hotel. Shortly after there was a fire at Hagan's Ranch where two buildings burned.

The biggest setback suffered by the stage company was in 1881 when an employee of the company contracted glanders from one of the company horses and expired. The widow brought suit against the company and collected a large amount of money and forced the company out of business on this line and to sell much of their equipment—also the other companies were crowding them out. A portion of the equipment was purchased by the Wyoming Stage Company that had been on the Pierre route, including the barns and stables along the route . . . and to add to this purchased six new eight-passenger coaches . . . with double curtains. Efforts were made to extend the line to Deadwood which were finally successful and the mail came directly to the gulch city; the Wyoming Stage Company had received the mail contract from Sidney to Rapid City. In 1884 they were running four daily coaches. In 1884 the Sidney stage route stops were listed as North Platte, Fort Robinson, Hot Springs, Buffalo Gap, Rapid City, Bear Butte, Fort Meade, Sturgis, Boulder Canyon and Deadwood. At the stations were good facilities and good meals were served.

By August 1, 1885 Sidney was abandoned as a terminus and Parker and Marsh did a large business from Chadron. Several towns were spawned by the railroad and took turns as the terminus, and a few remained after the rails pushed on. Some of these camps were Valentine, Gordon, Chadron, Pimpville (on the crossing of the Cheyenne) and Buffalo Gap. Chadron was the queen bee in 1885 with ten saloons, four hardware stores, two drug stores, five grocery stores, hotels and three furniture places. There were about two to three thousand people in the town. And of course there were the usual gambling facilities and faro and roulette flourished . . . it was a lively—and rough place.

The first Sidney coach via Chadron left Deadwood on July 31, 1885 and the first Northwestern coach arrived in Deadwood on August 21, 1885. The people in Deadwood gleefully noted that it took only twenty-five and one-half hours to get to the railroad—a decided improvement on the days it had taken days! Chadron's glory was brief and by September it was practically a deserted town with only one saloon remaining and Pimpville was in full bloom. Shortly Buffalo Gap came in the picture and was a busy railroad town for a number of years. In 1890 the railroad reached Deadwood . . . and overland stages to the Hills were past history.

Parker and Marsh transferred their operations to their lines in Wyoming, Montana, Nebraska and Utah. In 1896 the old Sidney Stage barn that stood in the alley facing the Bullock Hotel in Deadwood was being torn down . . . by then it was a dilapidated old building. And since the tourist season was approaching people thought it was a good idea to get rid of all the old eyesores. One of the stages was used between Custer and Sylvan Lake for a time—and unfortunately, for some reason or the other, sank in the lake, where it remained for five years. It was finally raised.

*Photo of Calamity Jane taken in 1880 when she was all decked out in her "glad rags." We usually think of her dressed in buckskins and toting a gun that she often toyed with and bragged, "When this dog barks somebody drops." However, that likely was talk—in saloon fights she much preferred to rush in and knock the fighters aside and after the fisticuffs were over waltz up to the bar for some strong potions.*

*Photo "The Huffman Pictures, Miles City, Montana."*

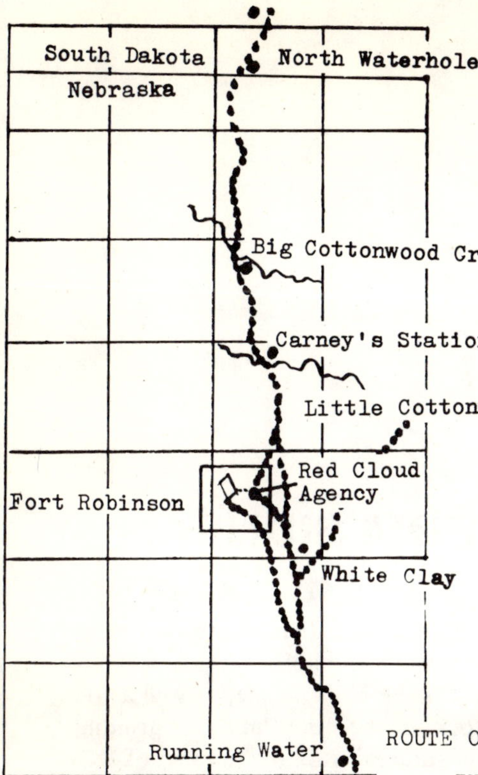


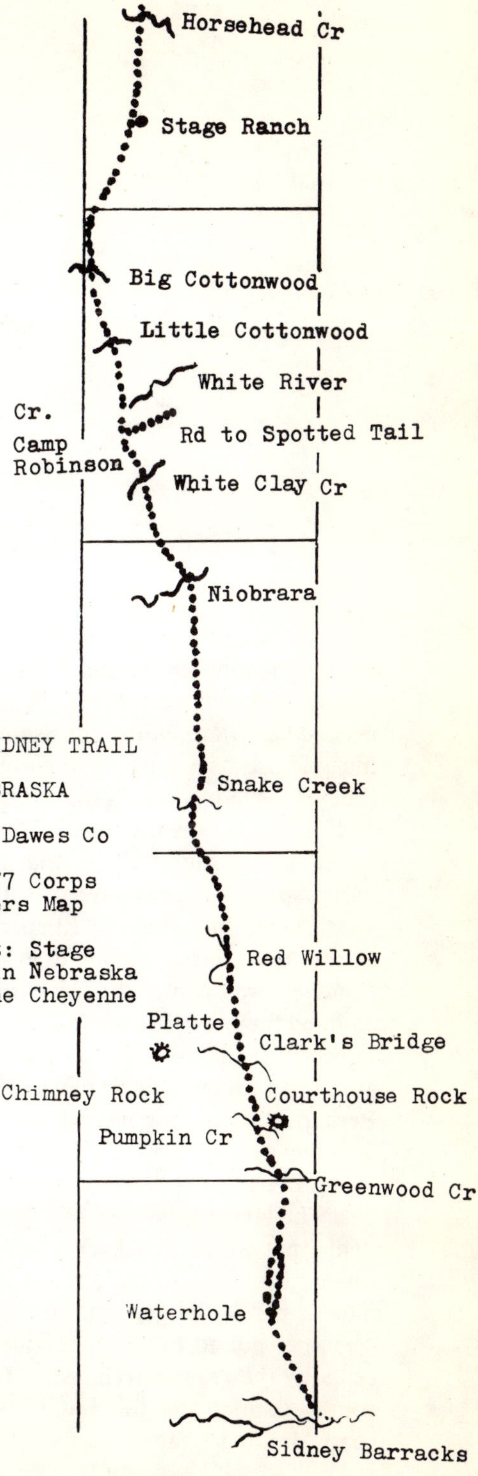


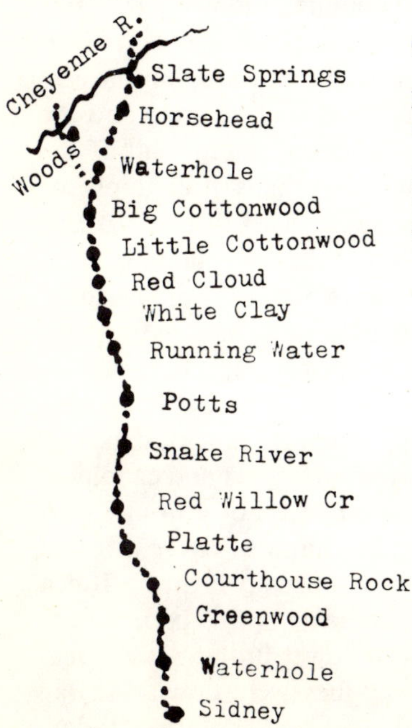


ROUTE OF SIDNEY TRAIL
IN NEBRASKA

Top left: Dawes Co

Right: 1877 Corps of Engineers Map

Lower Left: Stage stations in Nebraska & SD to the Cheyenne

# 5
# FREIGHTING FROM SIDNEY

The story of freighting is perhaps less romantic than the story of staging but the plodding teams of oxen and the mule and horse "fast freight" or "quick freight" teams brought in food for the Hills people, as well as the needed hardware, house furnishings, stock for the stores, mills for grinding the gold ores, sawmills for building homes and mills. In the summer of 1877 the Nebraska merchants were shipping 275,000 pounds a month to the Hills and in July it was noted that there were twenty sawmills in the Hills with seven more enroute. The people needed these necessities of life, but the early newspapers often referred to them as the luxuries and the beer and the whiskies as the necessities. Especially welcomed was a wagon load of the "brew that refreshes but does not intoxicate"—perhaps the supply of "budge" was depleted. The early Hills people were hard drinkers if one counted the breweries and saloons—in 1877 Gayville had two saloons and a brewery, even earlier the J. J. S. Brewery in Deadwood was going strong, and Lead when only three weeks old had two saloons and a brewery. Perhaps the drinking helped erase the hard realities of living, the loneliness of many who were away from their families—and the fact that gold nuggets weren't teeming in every panful. Not too many made that lucky strike.

The fast freight lines brought in sugar, flour, coffee, tea and candles—items that weren't bulky but very necessary. The owners of all outfits arranged for "hay stops", for supplies, repairs, etc., and to cover this and make a profit charged what the traffic would bear—usually eight to fifteen cents per pound by the quick freighters and two cents up to six and eight cents by the regular freighters. The charges were lower in the "easy" seasons—summer and fall but in winter when supplies were down and the roads in bad shape charges were more. As with the inflation today, freighting costs got higher and higher each year. In the very early days the freighters had been able to feed their animals by the wayside but this wasn't possible when there was more traffic.

There were individuals who made only a single trip and settled in the Hills, the freighting in of merchandise and carrying passengers helped pay expenses. One of these men was James Brown who came here in 1879. He started from Iowa and followed the railroad across much of Nebraska to Sidney, a place he described in a letter to his wife as "much of a business place and contains people of every hue and toning about." Here he and his brother, John, contracted for two hundred pounds of freight, seven hundred pounds of corn and five passengers with their luggage. The passengers were "one robust old Frenchman, one spicy Englishman and one burley Scotchman, one wide brim Texan and one Wapello County lad," all of whom were "armed for war fair, so the redskins did not show themselves and we had peace in the camp most of the time . . ." They arrived in Deadwood the early part of May 1879.

James went on to write that his passengers and freight brought him fifty dollars. "I had to pay $5.40 tole (toll) on the road and get my team shod and buy an axe and bucket and pear of overalls and a few other little things besides my traveling expenses and you know I had $21 when I started and I have $35 now . . ." He planned to buy a new suit of clothes so he could go to church and "not be asked to trade or work or gamble all day Sunday you have no idea what a rough set of people there is here." He did purchase a suit and posed for a portrait, looking very handsome indeed. Later he was short of money and regretted the expenditure but what was done was done—and his family members, no doubt, were happy to have a picture of their husband and father.

James eventuallly settled on Spring Creek north of Crook City but while in Deadwood did a spot of sightseeing. He went to Lead City and from there up the gulch to the head and "onward and upward at times onward one length and upward two until I ascended to the summit of Terries Peak* which is said to be the highest peak of the Black Hills." It was the first day of May and there were fields of beautiful flowers even if the canyons still held drifts of snow three to ten feet deep. James returned to Deadwood going past Bald Mountain to Silver City and then down the gulch "between the steep mountains on each side so close that the miserable road is made by diging off the hills and filling up the ditch between, done to anchor them down to Central City, down again to Gayville and finally three miles to Deadwood which is built over a creek, whose water is as red as the blood from the crushing of the red quartz by the mills at the towns mentioned above."

James noted that most of the land was staked with mining claims and worked in "maney places." He continued, "You have no idea what amount of interest there is here in the mines, the mines are worked over a territory fifty by fifty miles of gold baring mines with occasional silver and iron and coal oil and the finest izenglass that I ever saw, and the great green pines that cover the hills its more than I could describe . . ."

Brown may not have been the best speller but he described the area around Deadwood very well. And the mines with their gold and silver deposits brought in an ever increasing number of people and the freighting business flourished.

The small outfit that Brown brought in apparently hit good weather and fair roads from Sidney but earlier in the season a mule team loaded with batteries for the Homestake Mining Company hit mud and took thirty days to reach Lead. Time-wise the mules usually rated circles above the plodding oxen and one time a team of mules passed an oxen team that had already been fifteen days from Sidney. Merchants needing goods in a hurry patronized the mule fast freight and usually got their goods in jig-time.

Sidney prospered with all the freight and passenger business and expected to oustrip Cheyenne as soon as it got the stage business away from them. The Nebraska city was described—tongue in cheek—as a moral and a Sunday School village—with a number of murders and shootings. William Kane hammered William May almost to death, there was a bullion robbery and Jesse Donaldson and Banjo Bill had an altercation and Bill's head was beat into jelly. These two men were members of a hard gang that ran things in an evil manner.

The freighting companies included Pratt and Ferris who originally had carried supplies to the Indian agencies from Sidney and Cheyenne. There were other outfits including Daughterty, Kelley and Company, Jewett and Dickinson, Kennard and Simpson, McCann and Company and other smaller outfits. An outfit came in with machinery weighing 400,000 pounds —no doubt for the Homestake that was a good customer.

In 1879 the freight shipments from Sidney had fallen off considerably but received a boost after the Deadwood fire when a large portion of the camp with its wooden structures was wiped out. Immediately the energetic Deadwoodites started re-building, mostly fireproofs—and traffic was heavy from all directions. Ed Tomlin came in with 60,000 pounds of freight but had to camp on Centennial Prairie to wait out a storm. An eight mule team came in with 9,000 pounds of freight, another freighter with 106,000 pounds of machinery for the Caledonia. In January of 1880 J. M. Woods came in with a fine mule train, the mules all large and fat. J. Ellis and Company had good stock, too . . . their outfit came in with twenty-one teams consisting of 135 head of oxen that were the finest and largest ever seen in the Hills. By May of 1880 most of the companies had moved to the northern routes—it took about three weeks from Bismarck and even less from Fort Pierre. The Sidney and Black Hills Transportation Line owned by Pratt and Ferris passed to Jewett and Dickinson and later consolidated with the Merchants Line. By 1884 the Sidney Trail was a neglected route when the freight came to the Hills via Valentine and Pine Ridge that were at the head of the incoming railroad . . . and grass started growing on the Sidney trail.

*Harney Peak is the highest peak in South Dakota

# 6
# THE BRIGANDS AND THE HEROES

Gold shipments attracted the bandits as a rock concert attracts today's youngsters—in 1877 the newspaper deplored the fact that the Sidney coach had been held up AGAIN, this time by four masked men who stopped the stage this side of Battle Creek. The treasure box was empty but the road agents didn't go away empty-handed; they lifted the valuables carried by the passengers including three watches, some gold dust, all the valises and a set of dentist's tools. Hard telling what they did with the last item unless they personally knew a dentist willing to buy hot merchandise.

Shortly after five masked men held up the coach ten miles this side of French Creek and had no trouble opening the treasure box because they had a key selected from a ring of keys—pointing to an inside job.

In July of 1877 both the Cheyenne and Sidney coaches were robbed between Lightning Creek and the Cheyenne River by James E. (Lengthy) Johnson who was the terror of the stockmen in the Spearfish country. In 1880 three bandits held up the incoming and outgoing stages, two of the hold-up men wore dark clothes but one had a white hat—proving that it wasn't only the good guys that wore white! A robbery in 1877 involved an innocent passerby, or we should say two passersby when the Sidney coach was stopped between French Creek and Buffalo Gap. When the robbery was in progress a Mr. Jorgenson and partner drove up in a buggy and were promptly relieved of their gold sacks—sort of a bonus for the highwaymen.

The oddest robbery, without a doubt, was when the ropes that tied the treasure box to the roof of the stage were loosened and the box fell somewhere between Fort Meade and Sturgis. Later the box was recovered in Sturgis—empty as a freighter's billfold after a Saturday night "on the town." The box was again placed on top of the coach and unbelievable as it sounds, it was lost again before reaching Deadwood.

On and on went the stories of the depredations by the knights of the road including Jim McDonald, Fritz Stark, Lame Bradley, Frank Harris, Lame Johnny and Billy Mansfield. The stage company brass decided something had to be done to stop the brigands. They had the Master Safe and Lock Company of Cincinnati manufacture a burglar-proof safe that was "boss." It was outfitted with torpedoes and was perfectly safe if opened correctly but "worse than lightning" to green horns. Superintendent Voorhees and a blacksmith decided to see exactly how sturdy it was and found that they were able to smash it to smithereens in no time flat. Some more work was done on it and in August of 1879 when some robbers took it, they made no impression on it for twenty-four hours.

In 1879 the company treasure coach, Old Ironsides, was placed on the Sidney line. The Homestake gold was transported in this as well as the bullion from the other mines and it wasn't unusual to carry $200,000 in a single shipment. In early 1880 this amount was carried on the treasure coach the result of a two-weeks clean-up from the Homestake in Lead and the Father DeSmet in Deadwood Gulch. Another time five gold bars were shipped and duly noted in the newspaper—no wonder the road agents were more than mildly interested and they were right in there pitching after reading the local gazette.

In 1885 the stage was held up five miles below Running Water but the thieves might have saved themselves the trouble as the treasure box was empty. When the coach was running to the end of the tracks it was stopped by road agents, one of the last hold-ups. There was a single occupant in the stage, Richard Bullock, the messenger—and no treasure—the robbers were foiled once again. Bullock had the distinction of never losing a single ounce of gold during the years he was a messenger for the bullion shipments.

One of the big names on the wrong side of the law was Doc Middleton who made his first appearance in Sidney in 1877 where he killed a man. He jumped town by boarding the Deadwood Stage to Fort Robinson sans overcoat and it was mighty chilly weather. A man named J. E. North gave him a buffalo robe—which turned out to be a good investment as Doc never stole one of his horses—after he became a "dealer in stock." Because Doc was crazy about horses—other people's horses and made stealing them his career for a good many years. He stopped travelers occasionally, may have stopped a stagecoach or two but concentrated on horses. However, Doc never made illegal "acquisitions" in Holt County, Nebraska,—the reason being a buxom, dark-eyed lass named Pood Richardson who fell in love with Doc "on sight." The feeling was mutual and the two were married.

In 1879 Doc Middleton was described by a reporter for a Deadwood paper as being thirty years old, tall, slightly built and with a light complexion. He sported a black beard, moustache and had dark hazel eyes. His front teeth were filled with gold and he was often called the "man with the golden teeth." He carried two revolvers and was said to be a dead shot, capable of plugging a silver dollar in the air. His men both loved and feared him—and it was said the lawmen weren't anxious to get too cozy with him either. One time the outlaw stopped South Dakota's Senator Moody and divested him of everything but his jeans and then started him home for Deadwood, saying that the people there would miss him if he didn't show up.

Middleton was a master horse thief; one time he stole two thousand ponies from the Sioux which was quite a feat to accomplish with the prairie-wise redmen who themselves weren't above adding to their herd in the same manner. The outlaw wasn't even particularly reluctant to appropriate a few horses from the government. He got to be such a menace that in the summer of 1879 the people in northwestern Nebraska made a giant drive to capture Doc and his gang. A detail of soldiers under Detective Lleywllyn and the sheriff of Hall County got together and crossed the Niobrara to the home of Doc's father-in-law whom they captured and five others. Richardson was made to lead the way to the outlaw hangout situated in a deep canyon. Here the mounted men found Pood in camp, she was asked to get her husband to surrender but she refused. The men scouted around and found Doc crawling through the deep grass—and the master horse thief was in the hands of the law. The camp was destroyed as was another one nearby on which the gang had been working.

This was the beginning of the end for the rustlers who were hunted down. The newspaper reported that Doc's pony boys were shot as fast as they were found—this included Elmer Lewis, Sheldon and Davis who were found ten days later full of bullet holes. The last of the gang was arrested in Hot Springs, Arkansas, but the top man, Doc Middleton, became somewhat of a celebrity, a Wild West character. In 1888 it was noted that the sod house ten miles east of Hyannis was crumbling—the place that had been the rallying site of the robber band. In 1895 the "old-time terror of the Elkhorn" was in Omaha in Billy Carter's Saloon where he was on "exhibit for tenderfeet who had never seen a famous cowboy." The King of Horse-thieves lived in a number of places, in Chadron where he converted and joined a tour. In Gordon he had a saloon and the townspeople wondered why Doc had been allowed to build on a school section and another man in the town was not given a permit. Middleton visited Deadwood and seriously considered living in the gulch city—at the time he was described as a character of many a dime novel—the books young lads took to bed with them to read under the covers, when they should have been sleeping. At one time Doc was working on a Wild West show similar to the one Buffalo Bill Cody got together but it evidently didn't pan out.

There was another desperado closer to home . . . Lame Johnny, horse thief, road agent and bad man, with a reward on his head offered by the Sidney Stage Company. Johnny, too, was very fond of horses and wasn't particular whose ponies he took over. On one of his sorties to Pine Ridge he was arrested and on his way to Deadwood he and his guard were stopped by some armed men (it was thought one of them was Boone May), and the outlaw was hanged on the limb of a tree and the next day buried by some freighters. The nearby stream is known as Lame Johnny Creek. A headboard was erected on his grave with the following message: "Pilgrim, pause, You're standing on the mouldering bones of Limping John. Tread lightly, Stranger, upon this sod, For if you make a move, You're robbed, by God!"

Lame Johnny was born Cornelius Donahue in Philadelphia and as a small lad was thrown from his horse and as a result had a small limp for the rest of his life. He changed his name to John A. Hurley and went to Texas where he engaged in horse stealing, later he came to the Hills and for a time was a law-abiding citizen . . . a deputy sheriff in Custer County and a bookkeeper for the Homestake Mining Company. One day an acquaintance from Texas recognized him as a horse thief, Johnny quit his job and once again turned to being a "dealer in stock," as well as stopping a few stages.

One of Lame Johnny's last robberies was in January of 1879, (he was hanged the following July) according to Frank Harris, who testified in Rapid City in February of 1880. "At Dry Creek six miles this side of Buffalo Gap me and Lame Johnny robbed the coach about 4 pm. Stopped the coach and Tom Moore went through the passengers. Lame Johnny stood on the opposite side and made the passengers throw up hands, I hid in the bushes. We got two silver watches, one open face stem winder silver watch with a kind of flashy brass chain. Lame Johnny gave that watch to Goodwin, blacksmith at Custer, for grub. Moore kept gold watch, many two cent stamps and small change including two two-dollar bills, whiskey flask, small revolver, mail sack with bunch of stamps, several registered letters but no money. We left the sacks on the road and a pair of box-toed shoes. Later we camped on Beaver Creek, Johnny burned the stamps and got grub at Custer City."

The prosecutor wasn't impressed with the statement given by Harris, he believed the man was trying to shield Jim McDonald and Fritz Stark—and Johnny, of course, was long gone. McDonald and Stark had been accused of highway robbery.

People took ingenuous ways to outsmart the robbers. One gentleman hid a sizeable roll of bills in the fat between his shoulder blades and managed to pass through the "outlaw corridor" from the Cheyenne to Rapid City. In later years a hideout was discovered in the vicinity, and this evidently made it easy for the road agents to prey on the travelers going to the Hills.

Another canny individual was Barney McVey, an early comer to the Hills, who staked claims on Gold Run, located near the present Homestake Theatre. He later sold these claims to Sam McMaster of the Homestake Mining Company for $20,000.00. With a fat billfold Barney went to Ireland to get himself a bride and the two came back to the Hills via the Sidney route . . . with a good share of the money intact. The stage driver warned the passengers that holdups were frequent and not to carry too much money with them if at all possible. Barney checked out the stage for a good hiding place and then hid his wallet in a bag of oats at the back of the coach. Sure enough, road agents stopped the stage and stripped the passengers of their valuables and money—and thought it exceedingly odd that McVey was penniless. Barney just shrugged his shoulders and said he was a victim of hard luck he guessed. Mr. and Mrs. McVey stayed at Buffalo Gap for a week or so—which was no hardship for the couple as they had plenty of money for their needs.

There were "good guys" too, connected with the Sidney line—one of them was James M. Brown—not the Brown already mentioned. This Brown served as a shotgun messenger on the run. At one time when there was a fire in the stage office in Deadwood, Brown saved the valuables of the company that had been placed in the messenger's safe and let his own valuables worth about $1,250 burn. Brown took the safe to Gayville for safe keeping, carrying it all that distance on his back. When he ended his career of twenty-one years as messenger he started working for the Savage Mining Company at Pennington and in 1881 left for Tombstone.

Another worthy citizen was Henry Clay Vanatta who was in charge of the Sidney-Deadwood coach. In 1884 he drove a dray as did many of the old freighters and staging men.

Many of the personnel with the Cheyenne line later worked on the Sidney route and still later for the Fort Pierre outfits. All star members of the legion of heroes connected with the trails leading into Deadwood—the brave men who kept the wheels of commerce and transportation rolling to the Hills of Gold.

# 7
# ON THE TRAIL FROM SIDNEY

The Sidney Route changed as did the Cheyenne Trail to accommodate new camps and towns or to make a shorter or better route . . . however, it did not change as often as some of the other roads. Early accounts give varying mileages between the stations, many times this was due to an individual merely estimating the distances.

One-half mile west of Sidney the trail entered a small canyon with a gradual ascent for twelve miles to a plateau or mesa extending to Waterhole, sometimes called Dry Creek, twelve miles from Sidney. In 1876 it was a good stopping place for the wayfarers, the station operated by J. F. Hollowell. In 1883 it wasn't much of a place according to Edmond Mandat-Grancy, a French baron, who traveled the road to Sidney from a visit to the Black Hills. He told of a stop fourteen miles from the terminus—if this was Waterhole or another station we don't know because there is a discrepancy of two miles. At any rate the Frenchman was not impressed and described the barn as being one-half in ruins and at the station doorway was a mountain of tin cans. The room was small with a bed taking up two-thirds of the space, there was a stove, table and two or three stools, so there wasn't much room to spare. The horse tender and his wife were not prepossessing individuals . . they didn't even have one cow for milk, they didn't have a garden and no hens—quite unlike the thrifty French farmers and wives. The stock tender told the baron that they couldn't bother with raising anything—after all the company might transfer him at any time—and he certainly wasn't going to do all that work and have someone else reap the benefits. Besides the company brought in free bacon, potatoes and flour, so why bother with cabbage, beans and other garden truck or with eggs and milk when he and his wife got along on their present diet—and evidently didn't mind that it was a bit monotonous.

The road continued northwesterly for fourteen miles on excellent ground with a gradual descent to Greenwood Creek situated in a lovely valley with "stately elms," good buildings and corrals where the stage horses were changed. The water was from springs with clear, flowing water. One traveler didn't enjoy a meal he had here—he was gouged one dollar for a wretched repast of salt pork, bad butter and sour bread.

Pumpkin Creek was crossed in six miles, an excellent place for camping with a rapid stream of water and plenty of wood. Nearby was Courthouse Rock sturdily standing guard on the trail, and an excellent landmark.

The Platte River, about forty miles from Sidney, was an important milestone for the early travelers, the crossing made easy by a bridge constructed by H. T. Clark. Today the bridge is gone, only the pilings remain, but it lasted one quarter of a century. There was an earlier bridge that was burned by the Indians before it was completed and to avoid future troubles a guard house was constructed at one end of the second bridge. Pedestrians were charged fifty cents, a rider and horse one dollar, carriage and pair two dollars and a wagon with oxen seventeen dollars—rather a stiff toll but worth it considering the river that could be mighty treacherous.

About ten miles from the Platte was Willow Creek, sometimes called Red Willow, a most picturesque place. The water came out of a hill, forming a crystal pool and there also was plenty of wood. The ranch was operated by E. Morton and Company—meals were served at all hours and there were cigars and the best of liquors on sale. The stage stopped here twenty minutes for supper.

In October of 1880 the stage upset near the station, in fact the Sidney coach had several upsets, usually with no injuries but this time a couple of the passengers were bruised a bit.

In November of 1880 the Esmerelda, a mine in the Northern Hills, had some machinery left here—they had received five loads by oxen train, but needed the rest of it. They decided to load it on mule freight wagons that were much faster than the oxen trains—besides the mules often were used on winter trips and the oxen were put in pasture for the cold months.

Our French baron stopped here to eat and was surprised that the place had a hotel made of rough planks. Beef and mutton were on the menu, but neither were fit to eat—no wonder the Frenchman had such a poor opinion of American cooking.

From Red Willow the Sand Hills country was entered—the baron thought this less interesting country than the area to the north. Water was scarce and he went through a sandstorm that was a hair raiser, this evidently had a lot to do with his poor opinion of the vicinity. At the Sand Hills the trails divided with the main route shifting slightly to the north. In fourteen miles many of the trails converged at Snake River, called Blue Water in earlier days—certainly a more pleasing name. However, Snake River is descriptive of the stream that makes a serpentine course through the country. Here was a ranch operated by J. W. Dear that was "open day and night for the convenience of the traveling public." The place had a general store, furnishing goods, grain and provisions for the early-day pilgrims. Seven miles was the Point of Rocks, a famous landmark. Pott's Ranch was located here and nearby was Barrel Springs, a refreshing water stop for man and beast.

Eighteen miles from the Point of Rocks was Running Water—the Niobrara of today—where there was an easy crossing. Here were birds of many species with the game birds especially welcome to add variety to the menu of the early travelers. The accommodations were excellent with a ranch and general store, located north of the present Marsland. There is a marker at the site to mark the station located on land owned by John Hughes, the station about one and one-half miles from the present ranch home. The steel marker is one of several placed by the Crawford Historical Society to mark the stage stations in Dawes County.

In 1883 there was a cattle ranch on the banks of the Niobrara, a big outfit with two to three thousand animals—so said the French nobleman; however, he didn't mention a stage station.

The trail continued over Breakneck Hill, a hill that was mighty hard on the early travelers, although the area is packed with attractive scenery, bluffs and buttes, winding valleys and evergreens—the present main road is a bit to the west. The White Clay Station was located on the present Gene Thornton Ranch that has been continually occupied since the early days. From a nearby butte can be seen the goal of the gold seekers—the Black Hills, a thin, dark line on the distant horizon.

Seven miles from Breakneck Hill was the Red Cloud Agency (twenty-three hours from Sidney), where there were excellent hotel accommodations and three general stores "carrying large and complete stocks of merchandise." In early days twenty-three companies of United States troops were stationed at the agency.

In 1880 information was telegraphed to Deadwood giving the number of passengers on the stage—and sometimes even the names were given. Then if there would happen to be a VIP aboard the Deadwoodites could lay out the red carpet and shoot off sixteen cannon shots—or whatever they wished to honor visiting dignitaries.

In the early '70's plans were made to move the Red Cloud Agency from the North Platte—the original location—to the headwaters of the White River in Nebraska's Pine Ridge country, which was done in 1873. At first things were peaceable but soon tensions built up and in February of 1874 Lt. Levi H. Robinson was ambushed on Cottonwood Creek near Laramie Peak. Cavalry and infantry troops were sent in and set up a camp near the blockhouse of the agency, the tent camp named Camp Robinson in honor of the dead lieutenant. In May the camp was moved one and one-half miles west of the agency where permanent quarters were built; of the original buildings some still remain, disguised in white siding and used as tourist cabins. The fort was still in business in the 1940's when it was used as a base for training paratroopers. Today it is maintained as a historical site by the

Nebraska Historical Society.

In 1883 Edmond Mandat-Grancy mentioned the troops stationed here and he also mentioned the station where a woman rocked in a rocking chair at sixty miles per hour while her husband cooked up a prairie hen for dinner.

The original trail went east of the buttes that are between Fort Robinson and the present town of Crawford; however, if the freighters had cargo for the fort they followed a trail there.

The roads divided after leaving Red Cloud with the main one going north to Little Cottonwood that offered grass, wood and water; however, being only six miles from the agency many went on to the next station. In 1881 Mollie Johnson left her trunk at Little Cottonwood and Mike Haley, on a spree, stole the clothing from it. When the young fellow—maybe an old fellow—sobered up he felt so badly about his actions that he took strychnine, a heavy price to pay for the deed.

In the early days Squeaky Johnson lived near the trail and he worked out quite a system to get his letters on the stage—the drivers weren't allowed to stop just here and there, especially to pick up one letter. Squeaky took a long stick, notched it at one end, placed his letter in the notch and the stage driver could snatch it as the coach went by. Reminds one of the fast trains picking up the mail pouches from poles where the sacks were suspended.

The Little Cottonwood station was on land now owned by Frank Dahlheimer, the site marked by a sign placed by the Crawford Historical Society. No buildings remain at the station location—only weeds, pieces of glass and pottery, an excavation that apparently was a cellar mark the place.

Big Cottonwood, one of the best known of the stations, was ten miles further, located on the present Moody Ranch. It was on the bank of Big Cottonwood Creek that today is usually dry but in the early days noted as usually having water "around the bend." There was also a well with fairly potable water and wood and grass. The place was run by a man named Carney and was often called Carney's Ranch.

In May of 1882 two freight wagons burned near here and lost was $1,300 worth of merchandise—what kind was not mentioned in the newspaper account of the fire.

Land to the north our Frenchman called the Alkali Desert—the stage line went along the edge of Nebraska's Badlands. Mandat-Grancy was right about the alkali—the water is very alkaline; however, cattle can drink it unless it gets too strong in the dams. Today big pipes bring in good water from the south, it is run into stock tanks scattered here and there on the hilltops. This is golden country—at least in the autumn—pale grasses, stretching as far as the eye can see, covering rounded hills, with occasional cuts with chalky light colored soil exposed . . . good grazing country.

Four miles from Big Cottonwood and about seven to eight miles south of the present town of Ardmore are two buttes located on the west side of the present road. Near here was a fork in the road with one branch going to Hot Springs and the other to Horsehead . . . the latter to the northeast.

Twelve miles from Big Cottonwood was a prairie camp called North Waterhole or Farnam's (Farnum's) Ranch. In 1877 the place was noted as having a ranch in "full blast and everything the worthy pilgrim desires can be secured," built in 1876 by the Sidney and Black Hills Stage Company and purchased September 1878 by Gilmer and Salisbury.

Horsehead was the next stop reached on a good road over this practically treeless area although cottonwoods grow along most of the creeks. The ranch was "filled with all the delicacies of the season" plus having wood, water and good grass. This was an important station as both the Sidney and Cheyenne line came through here when the Cheyenne Route was changed from the Jenney Stockade road that was alive with road agents—the company figured there would be less hiding places for the bandits on the Nebraska road.

The place was managed by George Boland of Buffalo Gap who stayed in a two-room cabin, there was a horse barn and accommodations for the wrangler. This was not a regular meal stop but food was kept on hand for emergencies.

In the early days the next station was Buffalo Gap, north of the Cheyenne, but in later days there was a station at Slate Springs, a short distance (above five miles) from the Cheyenne. There also was a later-day station south of Horsehead known as Lone Wells, west of the present Oelrichs . . . on land now owned by Mr. and Mrs. Fred Fisher, in fact the station was but a short distance from their ranch home. There was a well at the site and an excavation showing the site of the barn. Avon Shaw of Oelrichs states that in 1927 when he was attending the Muhm School the pupils marked the Lone Wells and Slate Springs stations with wooden signs.

Slate Springs was a breakfast stop. In January of 1883 a passenger on the stage told of starting from Sturgis on Wednesday morning and reaching Slate Springs early Thursday morning where they had an early breakfast. The stage reached Lone Wells at 9:00 a.m. where they changed to an open wagon box on runners; however, at Big Cottonwood they were able to board the coach again and reached Willow Grove in time for lunch. The distance to Big Cottonwood from Lone Wells was given as thirty-five miles—another account gave the distance from Big Cottonwood to Willow Grove as seventeen miles.

The South Cheyenne was a shallow, easily fordable stream—except in rainy seasons when it would turn into a river, wide and deep. There were beautiful groves of timber on the banks and plenty of grass, making it one of the best camping places on the route. The freighters had a fireplace on the north side that they used for cooking their meals whenever they had to wait for the high waters to subside or when they simply decided to camp here for the night. Ten miles north was Buffalo Gap.

In October of 1882 Mr. Furlong, the superintendent of the Sidney Stage Line announced that the route would be changed to include Hot Springs fifteen miles distant from Buffalo Gap and intersecting the Sidney road at Lone Wells . . . by this route Slate Springs would be abandoned.

However, by summer the route went west at Buffalo Gap, on to Hot Springs and intersected the main road four miles north of Big Cottonwood. This of course, necessitated new stations—one was south of the Cheyenne River bridge on the Hot Springs route, about one-fourth mile east and south of the present bridge on land now belonging to Charles Hall. The building was situated on a flat area but traces of it are all gone.

The Baron told of the stage station at Cascade as in 1883 he stopped here to visit his friend, Gifford Parker, a soldier of fortune who was establishing a ranch in company with a couple of associates. Parker had talked the Sidney Stage Company to change their route to come by his town, from Uncle Sam the entrepreneur wangled a post office, he was building a warehouse with a store—and planned to get a hotel. There were some settlers living in eight to ten tents but all were planning to build more substantial homes when they got their families to move here. With all the big plans in the making, the station wasn't much to talk about—the Frenchman described it as a small building about five yards long located on the banks of Hot Brook. There were plenty of mosquitoes in the area—and in the station—and he didn't appreciate it when the Americans made a "smudge" that was supposed to drive out the mosquitoes—and humans, too, in the Baron's opinion. But he did give credit to American enterprise—building a town in the middle of nowhere by promoting it into existence.

The resort of Cascade became well known in the 1890's—the place situated two miles from Cascade Springs on Cascade Creek. In May of 1891 the place was booming—it boasted of a large hotel of one hundred rooms being constructed, there was a pavilion, a $30,000 stone bathhouse, a post office, bank, clubhouse, a store, bowling alleys, a livery barn and many residences. Cascade had hopes of attracting the railroad but lost out to luckier towns—and the town soon faded away. Today at the site is a modern farm surrounded by the old empty buildings when the two thousand people left the once prosperous town. Two families still live there—the Marc Lampheres who live in the old bank building and they rent out their modern home.

About seven miles to the north is Hot Springs which at one time claimed to be the greatest health resort in the states. Here are warm springs first discovered by the Indians who found the waters highly beneficial to their ailing people—and even the warring Sioux and Crow forgot their differences when members of the tribes came to "steep" in the warm waters.

Our French nobleman noted that in 1883 the place was beginning to build up—there was a chalet of "tolerably good appearance" with another nearby. Some ladies from Deadwood were here camping and looking after their horses, plus partaking of the waters. The Baron thought that camping outweighed any benefits they might derive from the waters.

According to Badger Clark in his delightful booklet, "When Hot Springs Was a Pup," the stage station was at about Sixth Street and River Avenue, with Dr. Stewart being the agent. The Deadwood coach reached here at ten in the evening and Sidney stage at two in the morning—not a very good connection. Later Chris Jensen had a stage line from Deadwood, to Custer and Hot Springs.

Many people believe there was another stage station—perhaps in later days in the Chamberlain home now occupied by Mrs. Lucille Hill, on Sheridan, with the stage road circling the house. The house itself is constructed of native lumber—fifteen inch wide boards put together with square nails. In front is a hitching post and the first step of the stairs to the porch was a sandstone stepping stone for dismounting horses. On the place was a good well—now covered.

Some freighters, travelers and local stages went to Pringle and then to Custer, in 1877 the Sidney stage joined the old trail at Buffalo Gap located on Beaver Creek about two miles northwest of the present town. George Boland had the stage station here as well as at Horsehead. In 1885 he leased lots on his land where the present town is located and charged twenty dollars for sixty months per lot. Soon there were twenty-eight business houses established—but there were complaints that no one had constructed a hotel. At that time the place was booming due to the railroad.

A syndicate was formed including Grier of the Homestake, Selbie of Northwestern and Dr. Dickinson of Lead who purchased one hundred and sixty acres for nine thousand dollars and then sold lots which went as fast as a Presley record. By November fifty thousand dollars worth of town lots had been sold and old Buffalo Gap disappeared from the scene with only one-half dozen buildings remaining.

The stage passengers—and later the people traveling by train commented that the water from Beaver Creek was clear but cathartic caused by the high lithium content. Good water was available at five cents a bucket and three dollars for a wagon load—but this water was only for drinking as the women still made coffee from creek water. Not only did the water leave something to be desired but wood was scarce; however, cowboys were not. They would come swooping into town as hawks, firing their guns and descend on the many, many saloons. There were a number of killings in the camp.

Buffalo Gap was a busy town when the railroad came there from the east and it became a freighting and staging center. Agno's Black Hills Express and Stage Company was busy traveling between the Gap and Deadwood and business was so brisk that he had to replace his light hacks with regular Concords. He might have over-extended because his stock was attached by creditors, and then too, Northwestern was giving him stiff competition. In March of 1886 it was noted that the roads were terrible and it took sixteen hours to reach Deadwood. Before long the rails reached Rapid City and Buffalo Gap quieted down.

In the late '70's the route was given from Amphibious Creek, evidently an early name for Beaver Creek, to Custer twenty-three miles away.

Thirteen miles from the Gap was a station on French Creek, the best station on the line where John Watson was the station keeper. It was a big loss to the company when the station burned in 1879 and had to be rebuilt.

In 1877 and 1878 a man named C. L. Edgerton freighted from Sidney to Rapid City with his favorite stopping place being French Creek—a lovely place then as now. There was a barn, post office, saloon, blacksmith shop all of which are still standing. The buildings had sod roofs, replaced many years ago and the upper part of the post office has been replaced. The barn was constructed of lumber cut with a whip saw; the post office of logs. In 1900 Edgerton moved on the place and he and his family lived for many years in the post office. In 1919 a new addition appeared on the ranch—a lovely, large house moved in from Rapid City and in which C. L.'s son, Charles, and family live at the present time.

Twelve miles north from French Creek was a station on Battle River about one and one-half miles southwest of the present Hermosa. The stop or hotel was a large building with ten rooms, a single story edifice that was part log and part frame—nothing is left of it now. The log barn was gone by 1908—the stage was around a bit longer and some of the local residents believe it was washed away in a flood as Battle River floods rather frequently.

The town of Battle River (later known as Strater) was laid out by Jack Hale an early comer to the Hills, who came in with a bull train in 1877. In 1878, 1879 and 1880 Jack brought in pigs—and on the last trip brought in some sheep. Later he owned a stock farm in Pleasant Valley, later he bought a ranch on the Belle Fourche and built a home near Devil's Tower.

In the early days there was a woman called Sage Hen who traveled with a freighter and stayed at Battle River when he went to Rapid City to unload his cargo. One day a couple young fellows asked her to fix them a meal and she agreed by whipping up some boiled bacon, coffee and lambs quarters, a type of greens. The men thought the greens looked rather unappetizing and declined to eat them. Sage Hen got her feathers up, took a six shooter from under her apron and informed her guests, "You will eat those greens and you will eat every bit of them and don't delay much about it either as I got no time to fool with smart alecks." The smart alecks meekly ate the greens and when she asked for a dollar apiece for the meal, they paid it without a single remonstrance.

Nine miles further was Spring Creek where in the early days it was noted that there was a stage ranch with plenty of water in pools, good grass and wood. In the late '70's the ranch was purchased by a man named Mackin who lived to go hunting much more than taking care of his hostel. A frequent visitor was Frank Hebert, a friend of Mac's, who often stayed to watch the place and help the stock tender.

Indians frequently passed the place and many times camped near a spring on the creek. (There also was a well on the ranch but it is covered over now.) There was quite a bit of trouble with the red men as well as with the road agents. One time during a first-class Indian scare the woman cook went to a safer place but Hebert stayed to see that fire was not set to the many hay stacks around the station. At one time a band of Indians descended on the place and "requisitioned" all of the groceries and canned goods, fortunately Mac had had the foresight to stash his whiskey in a hiding place, most prudently too, as the Indians were not able to find it.

Knights-of-the-Road kept things interesting, too—some of them were Lame Bradley, Lame Johnny and Fly Speck Billy to name a few. At one time Lame Bradley stopped at Mackens to hoist a few slugs of spirits but didn't have any money to pay for them. He offered to leave a log chain in payment which was not accepted. It wasn't long before the stage was held up—no doubt the work of Bradley and cohorts. Fly Speck Billy and Lame Johnny ended their lives as star attractions at a rope party but Bradley apparently escaped the noose.

At one time there was a toll road near the station—which wasn't a going concern too long. The freighters took a dim view of shelling out the required fee and made another trail to the east that passed over East Gap . . . which was fine except for one crusty freighter. This individual decided that the first trail over Black Gap was better and when the toll gate keeper asked for the going price the freighter declined to pay and backed his decision with a display of fisticuffs. The toll gate was eliminated.

Mac and the stage company officials had a falling out and the stop was moved about one mile to the west under new management. Mac then decided to sell his ranch and did so to a Mr. and Mrs. Francis McMahon who lived in the station building for a number of years until they built a new house across the road. This house built around 1880 is still standing and members of the McMahon family live there.

The town of Maverick grew at the site of the second station and prospered for a time. When the modern road was constructed the town moved to the south of McMahons. Ruts of the old trail can be seen about one and one-half miles off the present road where the wagons pulled out of the Spring Creek valley and went to Rapid City.

It was ten miles to Rapid River, today called Rapid Creek. In 1877 the citizens of the town surveyed a road between Rapid City and Buffalo Gap in order to entice the Sidney stage to use this route—and they were successful. (Before then the stage had gone to Custer from Buffalo Gap.) In August of 1876 there were two hundred settlers on Rapid Creek but the same month many of them left—some to the various mining camps in the Hills and others went back east with a train belonging to Louis Vollin. There was another exodus in September of 1878 when a large number of the inhabitants went to Rockerville, the largest camp in Pennington County—population four hundred.

North of Rapid City the stage stations changed locations frequently—in 1877 the first station was mentioned as being at Big Spring Creek and the next a ranch at Brook. In 1878 the stations were given as Hagans and then Crook City. In 1879 there was fire at Hagans and the company lost two houses. In 1883 horses were changed at Blackhawk and an 1884 map gives a stage station at NW 1/4, Section T3N-R6E. The early road approximated the railroad that came in later and wasn't too far from the present road. Stage Barn, about eleven miles north of Rapid City is said to have been a stop on the Sidney-Deadwood Stage line . . . and later it was on the Black Hills and Fort Pierre Railroad that started in Lead. Here is the Stage Barn Cave today—or perhaps we should say Cavern.

In 1877 the Sidney Stage route went north to Elk Creek that was "dry at crossing, wood plenty." There was a stage ranch over six miles further, that offered excellent water, grass and wood to the travelers—the station near the Tilford of today and the present highway rest stop, where there is an excellent spring west of the highway rest stop. Big Springs was the home base of one Mr. Morse—in fact Morseville was one-half mile west of Big Springs according to the early newspaper and Morse Creek is the stream that meanders through the area. When the Homestake railroad was in the planning stage Mr. Morse tried to persuade the company to build to his town but wasn't successful in his efforts.

The next stop, the Brook Ranch, was six miles from Big Springs somewhere near the present National Cemetery.

Eight miles further, a bit to the northwest (the trail circled the Hills), is said to have been a way station. In 1877 it was noted that here on Oak Grove Creek was water, grass and wood "good and abundant and near the Bismarck Road." Perhaps the first building was but a shack—likely with a sod roof—and later a more substantial building—and buildings were erected, if not by the stage company then by a farmer who settled here, as today can be seen depressions of several buildings. In earlier years there was a sturdy foundation of a house, evidence of a barn and a blacksmith shop and there was a good well to the south. Layton Terrill whose family now owns the land, said there are two mounds to the west of the place that could have been graves. Less than a mile to the west is the junction with the Bismarck Road that follows Spring Creek from the north.

Less than a mile from the junction is Whitewood Creek—the 1877 Engineers Report gave the statistics as: "stream 20 feet by eight inches, wood plenty, grass poor." Crook City was a little over one mile away . . . 1.38 miles to be exact. Crook City was one of the early camps to the Hills, located on Whitewood Creek that started in the high elevations and brought particles of shining gold down to the lowlands . . . there was considerable placer mining here in the early days. The town was an important place as the gate city to the mining camps and situated on the Fort Pierre, Bismarck and Sidney Trails . . . it had dreams of becoming a major trading center. Bull whackers, cowboys, gamblers, miners, merchants, all sorts of people clumped up the streets, visited the many saloons, stopped at the restaurants, did their shopping and other business—but by 1878 the camp had lost its glitter. The town had several revivals but in the end lost to Whitewood that came into existence with the railroad.

The stage trail turned toward southwest after leaving Crook City and in 4.85 miles made junction with the road to Spearfish and Fort McKinney, this near the present Duane Carstenson Ranch. It went past the ranch—that is the Deadwood Road, followed the ridge to the Thompson Half-Way

House, past the Rest (the present Preacher Smith Monument) and down Deadwood Hill to the mining camp. An earlier road was the Deadwood-Crook City Toll road and used for some years later that went further to the east and another that was further to the west and entered Deadwood by Centennial Avenue . . . this latter road called the Centennial Wagon Road . . . both discussed more fully in the chapter on toll roads.

Late in 1878 the route was changed to "accommodate the residents of the new military post on Bear Butte," this, of course, was Fort Meade built that year. Mention was made of a town called Dudley Town about one and one-half miles below Fort Meade and Rhulon on the "Sidney Road."

A cutoff road was made in 1878 that came up Boulder Canyon; this erased Crook City from the route. The stage company was considering a station at Boulder Park and may have changed horses here—there was a livery barn near the present day golf course.

In the summer of 1878 there was talk of having the main route miss Rapid City and leave Deadwood by Mountain Meadows, Camp Crook and Rockerville that would cut off twenty-five miles but this apparently was just talk. It was also considered having the road go from Hill City to Newton Forks, northwest of the town but again this may have never gone past the planning stage.

The freighters were independent souls—they followed a trail that suited them—usually the shortest. They did use the stage trails quite a bit—but we have a story of a freighter who came in when established trails were far between. This was Mahlon Williams who early in 1876 came up Battle Creek with a bull train and said that at that time there were no towns in that area. He went up a trail (a very skimpy one), leaving the present day Hermosa, up Battle Creek, passing the present day home of Mr. and Mrs. Virgil Williams, on to the north where the old Hattenback place was passed—the present day home of Mr. and Mrs. Chelsea Ferguson and not far from Rushmore Cave. In a field near the Ferguson home are piles of boulders placed there by the early freighters as markers to show the route and make it easier to follow, especially in the wintertime when snow was on the ground.

Mahlon went on to Deadwood and left the Hills by the eastern route to Fort Pierre. Soon other individuals came—and stayed—many of them in the Hayward, Harney and Rockerville area where mining claims were staked out by the dozens, gold was the theme song and riches the dream. For many the dreams of wealth did not materialize but when the gold fever ran rampant the freighter's wagons and stages answered the needs for transportation—for the people, for getting mining machinery and other supplies as well as taking out the gold dust. In this area of the Hills and in other sections.

The Sidney Route in company with all the routes that arched to the Hills played an important and exciting part in the days when rolling, rumbling wheels tied the gold camps to the outside world.

LET'S GO TO THE GOLD DIGGINGS
—and here are some of the listed routes

SIDNEY TELEGRAPH
January 6, 1877

Sidney
12
Waterhole
14
Greenwood
6
Courthouse Rock (Pumpkin)
2
Platte River
12
Red Willow
14
Snake Creek
18
Running Water
12
White Clay
6
Red Cloud
6
Little Cottonwood
10
Big Cottonwood
12
Waterhole
20
Horsehead
7
Cheyenne River
10
Buffalo Gap
20
Custer

CORPS OF ENGINEERS 1877

Camp Robinson
1.66
Red Cloud
7.21
Little Cottonwood
7.70
Big Cottonwood**
11.15
Dakota Line
10.15
Stage Ranch
10.72
Horsehead
9.20
Cheyenne River
8.02
Junction
1.05
Amphibious Creek
12.52
French Creek**
12.34
Battle Creek
9.33
Spring Creek**
10.08
Rapid City
1.31
Rapid Creek
12.68
Spring Valley
2.72
Elk Creek
2.06
Big Spring**
6.26
Brook Ranch
8.33
Oak Grove Creek
0.92
Junction
.096
Whitewood Creek
1.38
Crook City
4.85
Junction
4.88
Deadwood

Total Miles: 157

Later the route went around Fort Meade and through Boulder Canyon

**Stage stations

S. M. BOOTH
S. Dak. Historical Collections
Vol. XXV 1951

Sidney 12
Water Hole 12
Greenwood 12
Courthouse Rock 8
Platte River 8
Red Willow 10
Snake River 12
Potts Ranch 12
Running Water 14
White Clay Creek 6
Red Cloud 6
Little Cottonwood 8
Big Cottonwood 7
Junction 6
Cheyenne River 16
Point of Rocks 17
Custer 10

(Freighter's Route)

*Zenith Air Photo of Sidney—grown considerably since the 1870's and 1880's when the stage was coming to Deadwood. Sidney was a swinging place—at one time it had fifty-three saloons!—Courtesy photo: Nebraska State Historical Society, Lincoln*

*The Deadwood stage coach—usually drawn by horses but in this picture the power is supplied by mules. The stage line got a boost in April of 1877 when it got the mail contract and had it until 1878.—Photo courtesy Nebraska State Historical Society, Lincoln*

*On the right is Doc Middleton, notorious outlaw of the Black Hills and Nebraska region. Next to Middleton is Chippy Lamson, ranch owner and the round-up cook. Note the Dutch ovens in the foreground and part of an old stagecoach in the background.*

*Taken at Agate, Nebraska about 1900. Photo courtesy Nebraska State Historical Society, Lincoln.*

*Wedding photo of Doc Middleton and Rene Richardson, younger sister of Mary "Pood" Richardson, Doc's first wife.*

*W. A. Croley, Photographer, 1884. —Photo courtesy Nebraska State Historical Society, Lincoln.*

*Site of the White Clay Stage Station on Breakneck Hill that is east of the present road. This was a mighty hard stretch of road for the early travelers.*

*Here is a picture of Breakneck Hill—an area of buttes and valleys, evergreens and white sand—beautiful country.*

*Army wagon at Crawford City Park—a reminder of the days when the Red Cloud Agency and Fort Robinson were important places on the Sidney Trail. In December of 1876 the Sidney and Black Hills Stage was running three times a week with four horses between the agency and Sidney and once a week from the agency to Deadwood.*

*The chalk buttes between Crawford and Fort Robinson—sporting such interesting names as Devil's Coffin, Lover's Leap, etc. Another interesting butte is Crow Butte, southeast of Crawford, where the Sioux chased the Crows to the top—but in the night the Crows fled.*

*Bill Dodd pointing to an excavation at Little Cottonwood that may have been a cellar. Bill has done a lot of work with the Crawford Historical Society in marking the stage stops in Dawes County.*

*One of the signs put up by the Crawford Historical Society—this one at Little Cottonwood north of Crawford. Famous people came here on the Sidney stage, including Joseph Gossage who visited Deadwood in January of 1878. He did much to promote the Sidney route as editor of the Sidney Telegraph.*

*The stop was near here on Big Cottonwood—an important stop. There was less water in the creek than at Little Cottonwood, but in the early days it was noted that there usually was water "around the bend," the supply augumented with a well.*

*Two Buttes where the Sidney Trail branched, with one fork going toward the northeast and the other to the north to Hot Springs. Two Buttes were mistakenly noted as Ten Buttes in early descriptions of the trail.*

*Farnum's Ranch, also called North Waterhole, was just over the border into South Dakota. In later days it was known as Horsehead—in reality South Horsehead as the creek of that name curled from Cheyenne River way to the border.*

*The town of Ardmore in South Dakota, not far from the Nebraska border on the road to Hot Springs. At one time Doc Middleton and family lived in this area.*

*An old building at Cascade. This town gave serious competition to Hot Springs at one time as a health resort. The Woods Stage Station was to the south, across the Cheyenne.*

*This was Hot Springs at the turn of the century when it was a going concern as a leading health resort.—Photo courtesy Phoebe Apperson Hearst Free Library, Lead.*

*Sylvan Lake, a man-made lake, and one of the most beautiful in the Hills. The Sylvan Lake Hotel, the white building in the back, was a famous resort; when it burned it was replaced.—Photo courtesy Phoebe Apperson Hearst Free Library, Lead. From Meddaugh collection.*

*An old, old cabin on the wagon road that went up Battle Creek toward Hayward and Rockerville—this building on the Virgil Williams place.*

*The Harney Schoolhouse—now boarded up. Harney, Hayward and Rockerville (Rockville in early days), were important gold camps when the stages and freight wagons were the means of transportation.*

*The Point of Rocks near Pringle, an important landmark for the travelers going to Custer. The stage barn is still standing.*

*We can still ride a stagecoach—this one at the tourist town of Rockerville. The best coaches were the Concords, sturdy vehicles admirably suited to the roughest terrain. In September of 1878 "Rockville" had ten stores, one bank, seven saloons, a drugstore, four bakeries, three hotels and eight hundred people.*

*A sign erected by the Minnilusa Historical Society to mark the Buffalo Gap Stage Station, George C. Boland, proprietor. It was a short distance from present town.*

*The site where Lame Johnny was the star attraction at a necktie party . . . the original tree, said to be an elm, a short distance away. Johnny was a master horse thief but at one time was a deputy sheriff in Custer and later worked for the Homestake Mining Company as a bookkeeper until an acquaintance from Texas blew the whistle on him.*

*The old barn at French Creek, the lumber cut with a whipsaw. The doors were especially wide to allow the entrance of the freighter's oxen with their wide span of horns.*

*The "Main Street" at French Creek, now the ranch home of the Edgertons. The barn is at the right—to the left is the post office that was the home of the Edgertons for many years.*

*The post office at French Creek. The sides are of the original logs; the upper portion and the sod roof have been replaced.*

*Hermosa is but a small town today—although even in its heyday it wasn't a giant. The stage station was on Battle Creek a bit to the southwest of the town. The photo is of a building in Hermosa.*

*The site of the original station on Spring Creek. A man by the name of Mackin purchased it, when he had a falling out with the stage company the station was moved one mile to the west and placed under new management. The original stage stop was of logs, here one could purchase a few staples including Arbuckle Coffee. The second house—the one in the photo—was built in 1880-81 by Francis McMahon and said to be the only house of planed lumber on the Sidney Trail. —Photo courtesy Mrs. Tom McMahon and Mrs. Emmett Bennett.*

*Joseph Gossage who was editor of the Rapid City Journal after moving from Sidney.*

*The Spring Valley Ranch today—near here was the stage stop used by both the Sidney and Fort Pierre stage companies. The place was purchased by Captain C. V. Gardner who had a bank in the house in 1890 when Spring Valley was Piedmont. There was a stage station for a time at Stage Barn Canyon managed by George Nash.*

*The beginning of Fort Meade on Bear Butte Creek . . . this in 1878.*

*There was a way station here known as Oak Grove Creek, between Sturgis and Whitewood. Layton Terrill is walking toward the south of the stop where there was an excellent spring.*

*It is believed that this house was a stage station or a road ranch in Hot Springs—now the home of Mrs. Lucille Hill. Mrs. Hill is holding a hitching ring on a post.*

# 8
# FORT PIERRE AND THE EASTERN TRAILS

It was the first week of June 1876 when a large party of Black Hillers arrived in Sioux Falls with twenty thousand dollars worth of gold. The eyes of the local citizens popped at this display of the glittering dust and added several degrees to the gold fever that was prevailing in the towns, farms and hamlets of Dakota Territory. In the party were several freighters who had left their teams in Fort Pierre; with the men was J. B. Pearson, one of the first locaters of gold in Deadwood and Whitewood Gulches. He was enroute to Yankton, his home town, to persuade some of his fellow townsmen to go back with him to the new Eldorado. All of the men were full of stories of rich mining being done in the Northern Hills, of the many business opportunities for the ambitious; they were also muttering about President U. S. Grant closing the Fort Pierre route. However, it was soon opened again—first for supply trains and then for everyone and before long people from the east joined a flood of people from the other points of the compass to descend on Deadwood and the other mining camps to seek their fortunes.

The Fort Pierre route was the shortest way to the Hills, following old Indian pathways, the trail of the fur traders, early scientific expeditions and the military. In 1855 General William Harney had moved an army between Peno Springs, (also called Pinon Springs, Pino, Pinos, Pineau and Poeno) and Fort Pierre and stone fords had been constructed at the time across the creeks. Many a later traveler blessed these early engineers but had less than complimentary things to say about the clay soil that made traveling a nightmare in rainy seasons when the wagon wheels often sank up to the hubs in the gumbo and each turn of the wheel was a hard-won victory.

In February of 1877 the legislative assembly of Dakota Territory made a law to build three wagon roads, allowed by treaty, across the Indian Territory to the Hills; this was the Black Hills Wagon Road Law. One of the roads would be from Bismarck, another from Fort Pierre and the third from Brule City, then a spanking new town opposite the mouth of the White River. However, as these roads would traverse Indian country, Uncle Sam's blessing was required and the final decision was changed by lobbyists from Yankton and Custer City, looking out for their own interests. The Brule City road was changed to one starting from the Yankton crossing of the Missouri but this route was not widely used because Deadwood became the queen city and most of the traffic went there.

The Fort Pierre site offered three choices—Fort Pierre, Fort George and Chantier Creek, that was called Shanty Creek. The three locations were looked over by a committee—in fact two committees. The first one suggested Fort George, the second Chantier Creek but the freighters and travelers just went ahead and used their favorite, the one starting from Fort Pierre. And millions of pounds of merchandise was carried by the bull trains, by mule and horse wagon. Stage lines were born, flourished briefly and died—the most successful was the Northwestern Express, Stage and Transportation Company when they changed from the Bismarck Route.

Deadwood was emerging from a rough camp to an honest-to-goodness city. In the early spring of 1876 it had been the site of a forest but by April 28th a town site had been laid out. Buildings mushroomed and three sawmills within two and one-half miles of the town were producing 30,000 feet of lumber and 16,000 shingles per day—to make more buildings. Placer claims were producing 25 to 250 dollars per day and before long there would be quartz mining. Newspapers extolled the riches in the Hills bringing in more people and with an ever-increasing number of them via the eastern routes. The jump-off cities were Sioux City, Yankton, Fort Pierre and Chamberlain and for a time St. Louis was in the picture when they offered to deliver freight at four cents per one hundred pounds and passengers for ten dollars—and this included one hundred pounds of luggage. However, St. Louis was too distant from the Hills to give any serious competition to the other cities.

Travelers from Sioux City came overland past O'Neill, Nebraska, then veered right and went northwest to the Niobrara, to the Keyapaha, White River, Bad River, Cheyenne and to French Creek. Others went across Nebraska and entered the Hills from the south—two individuals who used this route—at least part way—were James Brown and William H. Frost whose stories are included. There were people who left Sioux City by train to Yankton and then took a boat from there upriver; some took a boat at Sioux City and didn't stop in Yankton. Persons arriving in Yankton usually took the boat to Fort Pierre but a few came via Niobrara, Keyapaha, Porcupine Creek and to Custer.

The Chamberlain route wasn't used widely because the Brules took a dim view of the white trespassers and weren't above lifting a few scalps or at least stealing a number of horses, oxen and mules. The route went toward the Badlands, following the divide between the White and Bad Rivers—much of the same route was followed later by the Milwaukee Railroad. Most of the Chamberlain traffic went upriver to Pierre and then across the prairies to the Hills.

In 1868 one rancher was living in Chamberlain, on American Creek but by 1880 the town started growing. In 1882 the railroad arrived and for twenty-four years stopped here. Chamberlain has the distinction of being the last steamboat town in South Dakota.

Two of the lesser known routes were by the way of Scotland, Olivet, Milltown, Rockport, Firesteel and Fort Thompson and the other was one used by the earliest prospectors that went on the divide between the Bad and Cheyenne Rivers to the mouth of the Belle Fourche and then on the divide between Elk Creek and to Box Elder.

Fort Pierre figured largely in the Montana gold stampede of the '60's as well as the gold rush to the Black Hills, but it has a history going back to 1843 when members of the Verendrye party buried a lead plate near the mouth of the Bad River claiming the land for the King of France. This plate was found in 1913 by some schoolchildren and is now housed in the historical museum in Pierre. In 1804 the Lewis and Clark Expedition camped nearby and in 1809, 1811 and 1812 Manuel Lisa was here and in 1817 Joseph LaFramboise built a trading post at the mouth of the Bad River near the present location of Fort Pierre. The post, called Fort Teton, was in business for five years until LaFramboise sold it to the Columbia Fur Company that built a fort at the same location. The American Fur Company later purchased the building and named it Fort Pierre in honor of Pierre Chouteau, Jr. who had an interest in the company. After the fur trade diminished Chouteau sold the place to the United States government who found it too small and in exceedingly bad condition and they built a new fort down the Missouri that they called Fort Randall.

By 1877 the old buildings and stockade of Fort Pierre had nearly disappeared; by then the camp of Fort Pierre a short distance down the river had a few log houses and a store run by a man named McGarry. Soon the place was growing as a "field of toadstools" and by 1879 there were three hotels, one restaurant, several saloons, a number of warehouses, blacksmith shops, several stores, one dance hall—with six girls and homes for the resident population of one hundred people.

The Fort Pierre residents weren't too happy with the growth of their rival across the river to the east that was beginning to have growing pains in 1879. The first house built in the later city limits was one constructed in the early part of 1877 by Napoleon Ducheneaux. In 1880 the train reached Pierre—where it stopped for many years—the Big Muddy was a mighty big river to cross—and it was the railroad that helped Pierre to outdistance her sister city on the west bank. By November of 1880 Pierre boasted of a population of five hundred, including three "legal lights," four M.D.'s and was considering hiring a schoolteacher as soon as a school was constructed. Of course there were the usual stores, homes, gambling houses and saloons and plenty of b-a-a-d girls.

In January of 1881 Fort Pierre had a setback when a fire destroyed several buildings including George Haines' store, LaPlant's dwelling, a drugstore, barber shop and a saloon. No sooner had the people recovered from the affects of the fire when there was a flood in March and a large part of the town was submerged. The residents were forced to the high bluffs where they watched the second calamity to fall upon their town. Floods in 1905, 1927 and 1952 again ravaged the town until ended with the construction of Oahe Dam.

In the early days flatboats took passengers across the river—a flatboat operator was Joe Kirley who lived about one-half mile down the river. In later days he ran a roadhouse west of Fort Pierre. Later steam ferries took the travelers across the river in the summertime and winter crossings were made over the ice. In 1883 the Pierre Stage Company had a small flatboat rowed by six men to take passengers across; they found it much faster than the ferry besides being capable of transporting two two-horse wagons with freight. Crossing on the ice wasn't always too safe and one time a ten-mule team and three wagons broke through and went in seven feet of water. The story had a happy ending because everyone—and everything—was fished out of the drink by Tom Douglass, a freighter. Another incident did not have an equally happy ending when a freighter by the name of Benway, who had consumed too much select lightning, fell from his wagon into the icy river where he was overnight. His hands froze solid and had to be amputated.

Ella Brown (later Mrs. Holst) stopped in Pierre in 1881 when she and her family were coming to the Hills from Iowa. She was the daughter of James Brown, mentioned in the Sidney and Bismarck stories who had come to the Hills in 1879 via the Sidney route. In 1881 he went by wagon to Pierre to meet his family, to join him on Spring Creek near Bear Butte where he had established a farm. Ella was a very young girl at the time and some of the trip was hazy in her memory but some incidents she remembered as long as she lived. Ella, her mother and her siblings rode the sternwheeler, the General Meade, upriver and it was a most pleasant trip although it was in July. The family took meals with the captain, his wife and sister and other agreeable folk who were also westward bound.

"Every so often," Ella wrote later, "the old boat would steam slowly down the river bank and someone would call out 'W-O-O-D P-I-L-E!' and the deckhands would make a run for the quickly placed gang-plank and begin carrying wood onto the boat from the many, many long ricks of cordwood placed close by to accommodate the steamboats."

The travelers "disembarked" at Fort Pierre, on the western shore and as it was Sunday they stayed with a family by the name of Doherty.

"Such a wind was blowing," wrote Mrs. Holst, "that when they finally sent the daughter of the house and me out to play a while, we had to wear cloaks and hoods, and one of our mothers had to open and close the door." In a couple of days or so the Browns started out and it took them five days to reach Sturgis. At nights they camped near water holes as did a number of freighters—but the children were not allowed to approach the freighters' camps. Fires were fueled with "gunny-sack" wood—buffalo chips gathered from the animal's bed-grounds, often the only fuel available on the prairies.

The children were continually told to be quiet so not to attract the attention of any red man who might be in the vicinity. Crossing the Cheyenne was a terrifying experience; it was deep and the banks were steep. "The water came just about to the bottom of the box," recalled Mrs. Holst, "and when the mules, Fan and Jule, started up the bank with the water running off their legs, they slipped to their knees sometimes and I was sure they would never get us out."

Sturgis had a deeply rutted street flanked by a handful of dirt shacks and slab buildings. One large structure housed the general store run by a man named Harmon, there was a busy blacksmith shop, the Ben Ash Stage Barn was a short distance to the northwest and across from the store was the Charles Hotel. Scattered about the camp were seven saloons—usually plentiful in the early camps—and Sturgis was no exception.

James Brown supplied his own transportation for his family as did many others, some booked passage with the mule passenger service and others rode the stage—if one was running. Not a few resorted to shank's mare and when there was Indian trouble often accompanied the freight trains—sometimes earning their passage by helping the freighters. Most wagon bosses charged a modest fee for carrying the baggage and food supplies of the travelers.

The stage took two days to reach Deadwood, mules took one week as did horse and wagons and the oxen trains fifteen days or more.

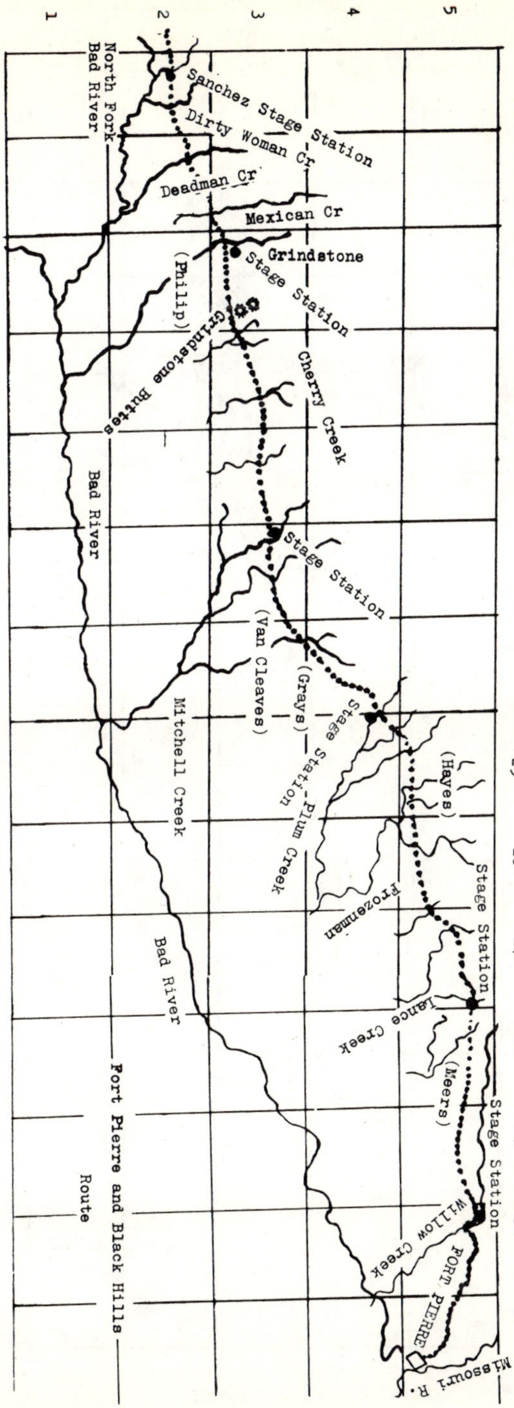

Sanchez Stage Station
Dirty Woman Cr
Deadman Cr
Mexican Cr
Grindstone
Stage Station
(Philip)
Grindstone Buttes
Cherry Creek
North Fork
Bad River
Bad River
Stage Station
(Van Cleaves)
(Grays)
Stage Station
Mitchell Creek
Plum Creek
(Hayes)
Frozeman
Stage Station
Lance Creek
(Meers)
Stage Station
Willow Creek
FORT PIERRE
Missouri R.
Bad River
Fort Pierre and Black Hills
Route

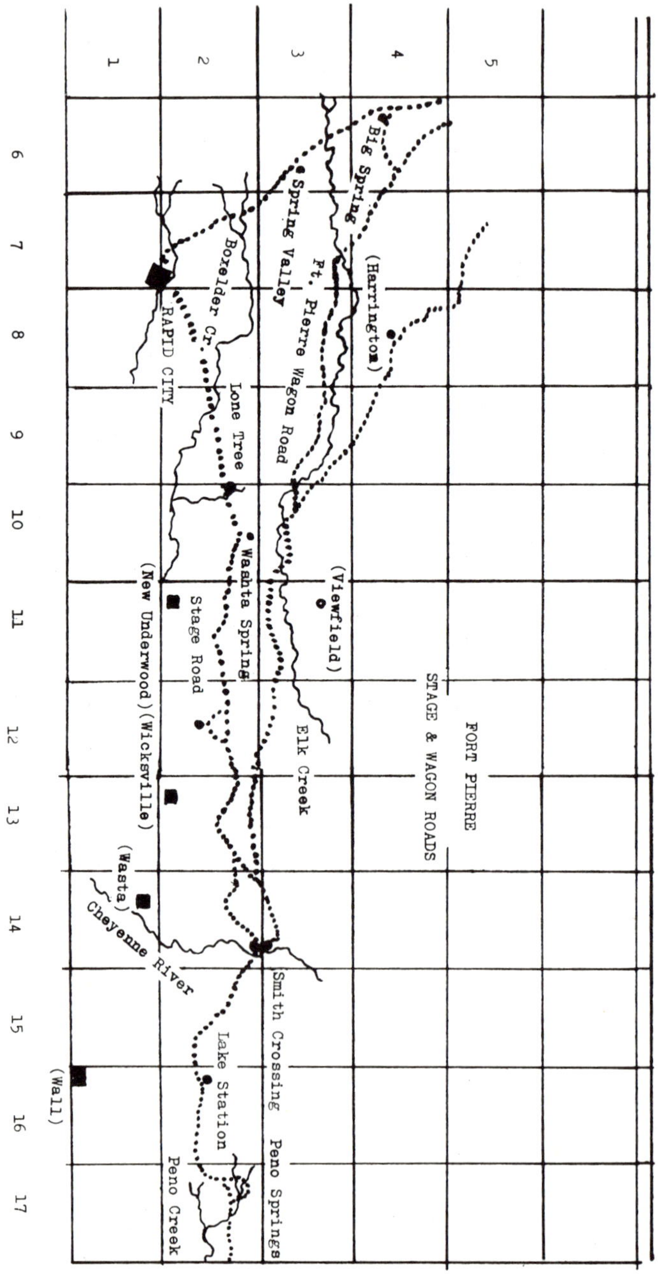
FORT PIERRE
STAGE & WAGON ROADS
Big Spring
(Harrington)
Spring Valley
Ft. Pierre Wagon Road
Boxelder Cr.
RAPID CITY
Lone Tree
Washta Spring
(Viewfield)
Stage Road
Elk Creek
(New Underwood)(Wicksville)
(Wasta)
Cheyenne River
Smith Crossing
Lake Station
Peno Springs
Peno Creek
(Wall)
1
2
3
4
5
6
7
8
9
10
11
12
13
14
15
16
17

# 9
# THE FORT PIERRE ROUTE STAGE TRAILS

Captain T. P. Phillips would have made a fine tour guide today taking tourists to Hawaii, the Far and Near East, to Japan or other parts of the world as in 1877 he successfully brought in a party of sixty-five individuals from Sioux City to the Hills. The newspaper noted that he was "fitted to conduct passengers without act or discontent,"—quite an accomplishment on a long, arduous trip of twenty-eight days traveling with oxen and mules. Under those adverse conditions tempers wore thin, patience was a forgotten virtue and quarrels and bickering frequently occurred between members of the traveling party. If the good captain was able to keep peace then—how easy it would be for him to conduct tours today with the improved transportation facilities now available.

Captain Phillips undoubtedly made only one trip—in those days it was a common thing to bring in pilgrims to finance a trip—in fact it was cheaper for the whole group.

If one had a fat billfold in 1876 he—or she—could book passage on a stage—one had been established in the early spring of 1876 by John Dillon and partner. This was from Yankton up to Fort Pierre and overland to Rapid City with the first trip made the first week of March 1876 with twenty-four passengers. Another trip was made June 11, 1876 when the stage was accompanied by twenty freight wagons, which may have been the last trip as the company was in business for only about a month.

Early in July of 1876 C. L. Demming contracted a stage line between Deadwood and Fort Pierre, headquartered in Gardner's Store. The first stage had five passengers and over five hundred letters from the Black Hillers to the outside world, the mail carried free-of-charge. The trip was under the charge of Captain Frank Hawxhurst, who on his return trip planned to bring all the mail that had accumulated in Fort Pierre and Yankton, he also planned to make regular trips, starting every Thursday from Deadwood.

This stage line either went out of business, combined or was the nucleus of the Black Hills Stage Company, also called the Fort Pierre Stage Line or the Fort Pierre and Black Hills that was in business for carrying passengers and express in April of 1877. The company was formed by Clarence Van Tassell, who had the mail contract between Fort Pierre and Deadwood, H. E. Adams (Anderson?), M. D. Gardner and a man named French. This was an ill-starred outfit beset from the beginning with misfortune, the first when a boat from St. Louis carrying company supplies of hay, grain and tents burned—with no insurance to cover the $2,000 loss. Some of the horses and stages had been sent overland and got to Fort Pierre okay but there was a wait for new equipment and supplies.

The Black Hills Daily Times said that the company received a number of fine, large coaches from the "manufactory" but Zack T. Sutley in his book, "Last Frontier" said the company purchased four old thorough-brace eighteen passenger overland stages. At any rate, the outfit finally was in business with stage drivers placed at the various stations. First class from Deadwood to Fort Pierre and then to Yankton by boat was $30.00—this included meals on the boat. Transportation to Chicago—by first class came to $49.25, second class to Yankton was ten dollars and to Chicago $26.65.

Business was brisk with every railroad train arriving in Sioux City or Yankton bringing in passengers bound for the Hills. The "live freight" as people were called, transferred to steamers that were "filled to the gills" of pilgrims going upriver. Yankton at this time claimed to be the leading outfitting point for the Hills, a wide awake town although it had suffered with four years of grasshoppers. Stories were of wild Deadwood, untamed Virginia City and other swinging frontier places but Yankton was no kitten purring on the hearth. It supported twenty-nine saloons and with the passage of years didn't get one bit more decorous as in 1885 it had more gambling places and vice than any other city in the territory. But the travelers weren't interested in a moral city—they were bound for the gold camps and many gave the wild life a whirl or two before starting out on the long trek.

In May of 1877 a passenger coming to the Hills described his odyssey, certainly a far cry from today's comfortable traveling in smooth riding cars, in fast planes and sturdy busses. The gentleman came by the Illinois Central to Sioux City where he made connections with the steamboat Florence Mayer that he described as a fine vessel. The fare to Fort Pierre was thirteen dollars and the trip took a week because strong winds delayed them. They passed four steamers, the Katie P. Kountze, Fannie Tatum, Kate Kenny and the Denver—all famous Missouri River steamboats. At Fort Pierre the passengers had to wait two days for the Fort Pierre and Black Hills Stage which finally arrived, a sturdy Concord drawn by four horses on good roads and two more hitched on when road conditions demanded, which was most of the time on this trip as the roads had no bottom due to the spring rains. Our passenger noted that the Cheyenne River was the only running stream on the route.

The arrangements for ranches had not been perfected and the travelers were two days without food. The stations listed were Willow Creek, ten miles from Fort Pierre, that had no water nor wood. Fifteen miles further was Lance Creek, ten miles Frozen Man's Fork, twelve miles to Cottonwood, fourteen miles to Mitchell Creek and thirteen miles to Buzzard Creek. Grindstone Butes (spelled with one 't') was reached after traveling twelve miles and another four miles brought them to Deadman's Creek—all these places were dry and with no wood. Finally after sixteen miles at Peno Springs there was water but no wood. The Lakes, nine miles further, had alkaline ponds and the people who drank the water became ill. The North Cheyenne River was eight miles and had alkaline ponds and no wood, neither did the Holes reached in five miles. Eighteen miles from the Holes was Wash Tay or Good Water Springs that had no wood but good water, in fact there still are a number of springs in the area. After Washtay Box Elder was approached that had water but no wood, then forty miles through the Red Valley, the finest country our traveler had ever seen. No doubt, he was more than happy to leave the unending prairies behind him and reach the Hills where there was water and wood—plus good restaurants offering hot meals.

The Allen Brothers "ranche" on the Cheyenne was one of the best stopping places on the route. It had been constructed in March of 1877 and had a hotel, stage stop and a store—a welcome stopping place for the wayfarer. Traveling in those days was no picnic, especially in the winter when icy winds shrilled across the prairies and the passengers huddled under buffalo robes in an effort to keep from being frozen and frequent stops were made where the people could warm up. Summer traveling wasn't a luxury either with the unbearable heat and the ever-present dust that mushroomed into billowing clouds and sifted into the coach leaving a film on the clothes, faces and baggage—and a gritty taste in the mouth.

The stage line was not a success financially. Adams and French went to the stations and took the horses—until stopped by Van Tassell and others. The case came to court, the bills against the stage company were allowed and paid and Van Tassell was able to keep the mail contract. However, with no stage the people who got off a Fort Pierre steamer either had to walk, hire a conveyance—or purchase one or take passage with a mule driver. In Deadwood in 1880 Gaston and O'Brien, stockbrokers, were named agents for the sale of coupon tickets for passage from there to Chicago, the Windy City with arrangements to Pierre by mule transportation, then to Sioux City and Pacific Railroad to the Missouri Valley Junction and from there the Northwestern Railroad transported the passengers on the last lap to Chicago in great style.

There was considerable scuttle butt of a stage line being placed on the line and finally in May of 1878 a tri-weekly stage was inaugurated; this company planned to make the trip in forty-eight hours for the one hundred and eight miles, with horses to be changed every fifteen miles. Stations were established at Rapid City, Washta Springs, the village of Smith, Deadman, Mitchell, Plum, Willow and Lance Creeks. The company hoped to get the mail contract away from the Sidney people but they weren't successful and the talk simmered down. In June of 1878 the mail was carried on small Indian ponies by the Dakota Mail Company to Rapid City and from there it was placed on the Sidney Stage for Deadwood. Soon the horses, hacks and the entire outfit of the stage line was hauled off to give the pony express a clear field—well, not quite clear because the young braves from Red Cloud's band were roving northeast of the Hills and were expected to form a junction with the Cheyennes—and both of them would give "attention to the rancher's hair" and, no doubt, to the hair of the pony express riders as well. There were several incidents with the Indians, one being in February of 1879 when the mail carrier and an employee of the Washta Springs Ranch executed a high dive into a hay stack when some red men started shooting and got one of their oxen. The beleagured men kept taking pot shots at the warriors who finally had to abandon the animal. Little by little the Indian depredations ceased and in June of 1880 nearly five hundred of them surrendered and turned in their ponies and guns. There still was trouble occasionally, especially on the Chamberlain route but the Pierre Trail was comparatively safe.

The talk of transferring the mail contract to the Fort Pierre line became a reality in July of 1881 when the Wyoming Stage Company transferred from the Sidney route. The company was on the Fort Pierre route for approximately eighteen months. In July of 1882 they sold their jerky line, complete with the mail and passenger service to Northwestern Express, Stage and Transportation Company.

The last of November in 1880 the railroad arrived in Pierre and when the port city was celebrating its arrival, the belle of the evening fell down some steps—none other than Calamity Jane, the buckskinned legend of many a dime novel. A short time later she went on to Chicago and the newspaper pleaded with the people there to "deal gently with this Black Hills exotic as she is a tender plant and could never endure rough treatment."

The railroad arriving in Pierre was the reason for Northwestern Stage moving their base of operations from Bismarck and they sent out Peter Seims to select stations between the Missouri River and Deadwood, followed by seven mechanics to construct stables, sheds and grain bins. The Indians weren't giving too much trouble but Captain Willard was sent out to give protection to the work crews—"just in case." By fall the stations were completed with the exception of the Elk Creek stop. The places had "magnificent buildings, first class with the best materials and built by number one mechanics." There were wells at all the locations so water wasn't a problem—no more drinking alkaline water from ponds. The company owned three of the eating stations with women presiding over the cooking pots and even the intermediate stops were large with good fires and water available.

There were sixteen hundred tons of "splendid" hay at the different stations, granaries were filled and firewood stocked. Someone jokingly said that "silver plated tableware is supplied and a load of French cooks has been ordered from Delmonicos." It was rumored that ice cream would be offered at the stops—which proved to be true! Fifty cents would be charged for most meals except at Rapid City where seventy-five cents would be charged. Jersey cows were purchased and delivered at the stations and even Deacon Selbie ordered some for Pennington where Northwestern kept their animals. The good deacon planned to use the milk produced by the cows for the wood choppers of his Sunday School classes—and the newspaper added that now little children were also attending the school.

On November 2, 1880 the first stage left Deadwood; however, there were no passengers aboard as this was a trial run. Everything worked out as smoothly as whipped cream and the line was in business. Big business too, as it carried an average of five thousand passengers a year, it eventually got the eastern and northern mail and most of the gold bullion shipped out by the Homestake, Caledonia and the other mining companies.

We are fortunate in having two accounts of people who left Deadwood by Northwestern enroute to the east. The first was in November of 1880 when the passenger boarded the coach in the gulch city at 7:30 a.m. and at 3 p.m. arrived in Rapid City. Perhaps they stopped at Brennan's place to eat, although our traveler didn't say; however, earlier in the year it was reported that "fat boniface Brennan is working on a way-up house in order to have it ready when the Bismarck coach line is transferred to Pierre." The stage reached the Cheyenne River at 11:00 where they ate; breakfast was at Medicine Creek at 8:00 a.m. where the weary passengers "scoured from their faces less than a peck of the sombre soil of the plains." Breakfast cost forty cents. At 4:00 p.m. they had supper at the Lance Creek Station and arrived in Pierre at 8:30 p.m. and crossed the Missouri on a steam ferry to make connections with the railroad. The stage driver on this trip was Scott Wickard.

The second account was by a passenger who left Deadwood the following month at 4:30 a.m.—the departure time changed frequently in order to make connections at Pierre and depended on the train schedules, the weather and road conditions. The driver "handling the ribbons" was Hank Williams, another well-known stage driver. The road was good, there were plenty of robes in the coach. At Boulder Park, a good-sized settlement, they picked up two more passengers. A good, hot breakfast was enjoyed at Sturgis City and a splendid dinner at Rapid City. The Cheyenne Station (where the company planned to build a fourteen by thirty foot ferry the next season—although it is doubtful that they did) was reached that evening and West Pierre at 4:00 p.m. the next day. The Missouri River had frozen over and the crossing was made over ice—no need of a steam ferry on this trip.

In 1884 during the summer and fall months the coaches made daily runs and to Chicago was only seventy-seven hours. Immigration hit full stride in 1885 with more and more people coming to the Hills to join the 40,000 people who had already settled here. If one stage could not carry the passengers waiting for transportation, another coach would be sent out—and sometimes even three coaches were needed. When the railroad approached the Hills, Northwestern started bringing in the mail, freight and passengers from the end of the track, which changed frequently. In 1886 the company got handsomely upholstered tally-ho coaches to run between Deadwood and the end of the track and planned to use six horses for the entire trip. Deadwood residents marveled that it took only seven hours from their city to reach the railroad—a fraction of the time it had taken in bygone days.

The terminus of the railroad changed from Valentine to Chadron, to Buffalo Gap and then Rapid City—with several small railroad camps in between. Then it reached Sturgis and on November 3, 1888 puffed into Whitewood. Northwestern Stage had not used the Crook City route for many years, but now once again—full circle—they started running daily coaches from Whitewood to Deadwood. In April they discontinued the Sturgis line altogether.

In February of 1888 Northwestern's mail contract was discontinued when postal cars started running through to Whitewood. Wells Fargo received the Homestake contract for the gold shipments and several hack lines started transporting passengers between Deadwood and Whitewood and Sturgis. Bartlett and Cooper took over the Northwestern run to Whitewood. The Northwestern barns were torn down in Deadwood and in 1890 the old barns and repair shops burned in Pierre. The newspaper reported it with a touch of nostalgia because it was there that Buffalo Bill had first fitted up his Wild West Show and where he had practiced. Also, the site had been the headquarters of General Custer and General Crook.

Hacks doing a profitable business at this time were those of Frank Smith who ran a double hack line between Sturgis and Deadwood, Jewett who had a line between Deadwood and Whitewood and Alexander had one between Lead and Deadwood. Tom Goodman had a daily line between Carbonate and Deadwood and J. H. Harris was running between Sturgis and Deadwood—and there were a number of others . . . both hacks and stages. But in time they all gave way to the horseless carriage that came rattling in with the 20th century and changed the entire transportation system.

# 10
# THE STARS OF THE TRAIL

The stars of the early trails were the staging men, the drivers, outriders, shotgun messengers. The stock tenders, the jehus or mule skinners. The operators of the various stops where fears of Indian raids and road agents, prairie fires and winter storms were ever-present. The bull whackers who drove the plodding oxen across the lonely prairies although in later years there were so many outfits on the trail that people described the incoming roads to Deadwood as one long string of wagons. There also was an army of pilgrims coming to settle in the Hills, with an average of fifty arriving in the gulch city every day . . . this was in 1886. The trip across the vast land wasn't a cream puff trip by any means but these early people were rugged, hardy and independent individuals for the most and who played a big part in settling the frontier west.

Some of the characters on the trails played the role of the "heavys" . . . the guys with the black hats. They were the brigands, the road agents and horse thieves whose real names are often forgotten and who sported such flamboyant monikers as Arkansas John, Curley Wood, Fritz Stark, plus several "Docs," including Doc Parsons, Doc Flick and Doc Middleton. Others were Van Moore, Frank Toole, Jim McDonald, Raymond Decker, Charles Hughes and J. T. Babcock to name only a few. The long arm of the law caught up with most of them, providing vigilantes didn't get them first—they had a tendency to arrange necktie parties for the outlaws without benefit of a trial. However, if the law ran its course many of the bad men served time including Babcock, called an old Black Hills Stage robber by the press, who served nine years in the penitentiary. Doc Flick served time and so did Charles Hughes, whose father had served with General Custer. Hughes also had to pay a fine of one thousand dollars. Van Moore was a horse thief—he was arrested in Pierre on a charge of getting farmers' horses from that area and then disposing them in the Hills. Bill Long, "Modoc Bill," was a rough hombre who was partial to horses—other people's horses that is. One time he was tailed by the lawmen to Castle Creek but they couldn't find him and then decided to spend the night in a cabin there. During the night Modoc entered the building when the men were sleeping, stole a rifle and before leaving wrote a note to his would-be captors, saying that he regretted not finding the cartridge belt to go with the gun he had "borrowed."

Texas Charley was a tough cookie, too. When an officer tried to arrest him at the Cheyenne Crossing on the Pierre route, Charley didn't favor the idea one bit and threw the lawman in the river.

These highwaymen appeared on the trails almost as soon as did the freighters and stagers; an early account tells of Louis Vollin robbed of seven hundred dollars on his way to Yankton from the Hills in February of 1877. Soon the free-booters were doing great—from their point of view—and the freighters had to pool together for protection from them as well as from the Indians. The vigilantes became active also and sometimes in their zeal would hang an innocent man as in 1877 when a group of men hanged a youth near Rapid City. The young man begged for his life and said he had put up with the brigands for company but the lynching party, caught up in the hot fire of the event didn't take time to approve or disapprove this and the unlucky youth decorated a hanging tree with his partners of the trail. However, in checking the youth's reputation in later years, it was learned that he wasn't altogether "pure."

In 1879 there were large raids of stock in the Northern Hills when it was reported that twenty horses had been stolen in Sturgis and over one hundred and fifty head on Redwater and at Spearfish. Doc Middleton and his gang of prairie bandits were suspected.

In 1880 a well-organized gang of horse thieves, with headquarters in Pierre, robbed an eighteen mule train at the Lake Station that was fifteen miles east of the Cheyenne River. They took all the provisions, plus two company horses and made their getaway. The same month A. Maupin was robbed a few miles west of the Mitchell Creek Station and a short time later every horse was stolen from a barn on the Cheyenne.

Stealing horses was a big business for many. Frontier people took a mighty dim view of these thieves and considered them the lowest form of humanity . . . and as with the youth and his companions near Rapid City, these individuals often ended their life at the end of a rope—including one near Spearfish.

Stages were lucrative targets for the road agents and one of the first stage robberies on the Fort Pierre route was in August of 1877 when the coach was stopped a few miles west of the Cheyenne Crossing at the Holes. These were extremely polite knights-of-the-road who asked for money only in excess of three hundred dollars per passenger and since none of them had even near this amount they were "excused from making a contribution."

Next the robbers ordered the baggage thrown off the coach which they examined closely but only two things were to their liking . . . a fancy shirt and a rifle. The outlaws were evidentally looking for big money and undoubtedly had heard that several well-heeled tycoons were to be on the stage. However, the wily capitalists had changed their minds and left the day before . . . one of them had been D. T. Bramble.

The top echelon men were the stage drivers, the men who "handled the ribbons", who could drive four to six horses with consumate skill. Some of them were Hank Williams, Jim Levelle—considered to be the best—Frank Ellis and Dan Callahan. Equally glamorous were the outriders who rode in front, sides and back of the coaches, especially the treasure coaches, to protect them from attack on the road by agents or Indians. There were shotgun riders perched on the seat with the driver, who fondled a shotgun and with a wary eye out for danger. Some of these individuals were Boone May, Billy Sample, Galen Hill, Jesse Brown, Gene Decker and John Cochran.

The men who guided the mule and oxen teams were heroes also—hardy and tough men who contended with the dust in hot summers and gumbo when rains turned the land into giant mires. Who fought blizzards, tornadoes, cloud bursts, floods and missing bridges—as did the stage people. Mudholes contributed a major headache to all travelers, including one in front of Fargo's Store in Elizabethtown said to be the worst pothole this side of Pierre. There was a wicked man-trap in front of the Minnesota Bakery on Sherman Street—old Frenchy fell in it on Thanksgiving Day 1881 and received mortal injuries. The bull whackers had to contend with broken wagons on the trail, including wheels that would become loose in dry weather or with heavy loads—and sometimes due to a combination of the two. They would then build giant fires and set the wheels.

Not only was there contrary weather and road conditions but there were contrary animals—especially mules—and some smart ones, too. The men would laugh about the mule on the Missouri that knew the recall on the bugle and whenever it sounded, quit work then and there and gallop full speed to the barns. They also chuckled about the old oxen in Deadwood that lay in the middle of the street for a little siesta and along came an express driver, who instead of going around the animal, drove over its legs. The bull got up and threw wagon and smooth Eph into the creek . . . which all agreed served him right.

An account in the Black Hills Daily Times told of a large mule train as long as a "moral law" belonging to Fred Evans that came into town. The mules were in good condition and the whackers "graduates of Yale—tall, muscular, aethestic looking chaps with nary a cuss word." Because soft spoken they were not, they could and did swear at sixty miles per hour, creating a blue haze about their heads. One freighter, Charley Barrett, was noted as speaking three languages—Sioux, Dutch and bull-whacker. However, there was one bullwhacker who arrived in Deadwood and used no profanity whatsoever and was able to handle a bullwhip with great dexterity—a lady bullwhacker.

When the hard-working freighters arrived from their trips, they would let their hair down a bit and the "whistling and melodious voice of the mule skinner" would change to whooping and yelling and the brandishing of pistols. "The dance houses would howl, brimming fluid flow festive, the dance red hot and joy unconfined." In 1878 there was one dance hall in Deadwood that was largely patronized by the whackers who "when the weird-like strains of the fiddle begin to float in the air, would snatch partners and bounce through changes in dance like spavined nightmares through a dyspeptic man's steeple dome of that during his slumbers." Journalism in those days was colorful to say the least.

At one time some bullwhackers exchanged shots at the Gem Dance Hall and one man was arrested. The newspaper scolded them saying, "These smooth Ephs will understand probably in the course of time that they cannot shoot their little guns in town with the same impunity that they can on the prairie."

Not that the men were quiet on the prairies—one time men attached to Clay's bull train got into a fight at Deadman's Run and a Mexican was killed. Also, a night herder belonging to Isaac's train was run over with a wagon while asleep. Sometimes there was a bit of celebrating on the trail—not all of the wild times were saved for the end of the journey. It is said that when the freighters came in via Boulder Canyon they would stop overnight near a roadhouse where the Kelly home is now located—not far from the golf course, and do a little syphoning from the barrels of brew that they invariably carried. A preliminary bit of celebrating on the completion (almost) of another long, hard trip across the prairies.

In 1882 Cheyenne Johnny arrived in the gulch city with a freighting outfit and proceeded to fill up on sheep dip and neglected his mules and wagon. The town marshall cornered Johnny and the two had a discussion on this—and Johnny ended in the cooler and his animals in a corral. The freighter was soon released but he had to "bail" out his mules for the time they had been corraled and for the feed they had consumed. The released man shook his head and marveled, "I am out on my own recognizance whoile me mules are kept locked up and shure they did nothing as they were sober all the toime."

Not only did the saloons have the freighters leaning against the bar drinking pressed corn but they offered several games of chance. The keno games used black buttons but the bullwhackers and rounders carried them off by the handsful and finally the managers were forced to replace them with pieces of glass.

The freight trains rarely ran in the winter time when bad weather set in—and the boats couldn't sail up the Missouri. However, winter trips were more common when the trains started running to Pierre and Chamberlain but even then the cattle were put to winter pasture when the weather became super bad. The site selected had much to do with how many cattle survived the winter. In 1884 Northwestern placed their oxen on Elk Creek and fed them hay and oats to fatten them up and took their most puny animals to

Pierre to give them T.L.C. The Bad River was often selected as a wintering area.

The freighters who played roulette with the winter weather and continued bringing in merchandise sometimes did handsomely—especially if the freight they brought in was in short supply. Sometimes, however, they would be caught in winter storms as in January of 1880 when a number of teams were caught at the Cheyenne Crossing where snow was over a foot deep. The men had to drag the ground with tree tops to get feed for the stock. Foley and McHume had already gone to winter quarters but Bob Hercher lost thirty head of cattle and Charley Dyer's outfit was stranded somewhere along the route. The newspaper commented that perhaps now the freighters would put up hay along the route the following season.

In the early days of freighting there weren't too many farms and ranchers on the Fort Pierre route as it was still Indian land but eventually a few ranchers moved in. One was Zack T. Sutley who purchased Tom Madden's Ranch on the Cheyenne and in company with a man named Smith who located about one-half mile down river put up hay that they sold to the freighters. Each party furnished one man and one team and were able to stack about one hundred tons—they also purchased hay from the other ranchers at five dollars per ton. It was quite a chore—especially when one day the thermometer registered one hundred and twenty-two degrees . . . but worth it financially.

The summers were hot on the prairies and the winters were cold—one of the worst experienced was the winter of 1879 and 1880. On the 15th of October of 1879 the first fierce storm swept through and lasted twenty-four hours. This storm was the forerunner of a series of storms that heralded a rigorous winter that brought hardships not equaled in the northwest. No thorough trains were able to get to the shipping points for months. The railroad cuts were opened but almost before the last drift was cleared, another storm howled in and furiously filled in the cut again. Steamboats were caught in the ice, the people suffered on their claims where they lived in shanties that they had expected to winter-proof in the usual moderate autumn. Food supplies dwindled to dangerous lows, coal supplies were exhausted and the engines had to burn corn for power on their extremely limited runs. Stages were lost and had to return to their starting points. The prairies were covered with two to four feet of snow—hard packed snow—that even made it rather hard going for the prairie-wise buffalo. Thousands of cattle perished, livestock as well as oxen used in freighting.

The thermometer dipped to record lows—twenty to thirty below recordings were not uncommon. Two men, John Brown and Thomas Kehoe started from Fort Pierre with a train—they were given permission to accompany it in return for helping the drivers. At the Cheyenne Crossing they stopped at Smith's Ranch and on December 18th started out again. The two men walked ahead of the train and took a wrong road—the one that went to Rapid City instead of following the freight road. They went as far as Washta Springs, a distance of twenty-eight miles. There was no protection except an open hut where they stayed all night and stamped about to keep from freezing. They must have realized that they took the wrong road when the train didn't catch up with them and in the morning they back-tracked. When they reached the outfit they were given a pony to take them back to the Smiths, with Kehoe mounted and Brown walking. At night they arrived at the ranch, Kehoe with frozen hands, nose and ear and Brown with frozen toes. They remained at the ranch for five days until an ambulance arrived to pick up a sick soldier who had been left at the ranch. Kehoe and the soldier came to Ford Meade and Brown came in later with some soldiers who were escorting the paymaster. When John arrived at the fort he asked to see his partner and was shown a grave in the distance—his friend had expired and had been buried. The military personnel were compassionate but said they could not keep Brown at the fort. In the morning he crawled to Sturgis where he was taken in by Johnny Scotland. Eventually Brown's badly frozen toes mortified and sloughed off.

Experiences like this were multiplied many times in that long ago winter but Mother Nature wasn't through—not by a long shot. The rigorous winter ended earlier than anyone could remember and the upper Missouri River and the Black Hills streams sent torrents of water down the Missouri. Near Fort Pierre the thirty or more inches of ice hadn't broken up and the floods swept on top of it taking bridges, both railroad and wagon, sweeping up livestock, buildings and household furniture. People were shut off from the outside world and many were on the verge of starvation. Then the ice started breaking up near Fort Pierre, Vermillion and Yankton and enormous ice gorges and ice packs formed. Raging torrents of water turned the rivers into icy seas and ice blocks carried boats far inland at Yankton. The river changed at Vermillion and farms that had been in South Dakota woke up to find themselves on the other side—in Nebraska. The flood destroyed hundreds and hundreds of buildings that had escaped the first flood. Buildings floated down toward New Orleans as so many majestic dowagers and others that remained in place were filled with silt. Flood conditions remained over a large area until early summer. The first thorough train arrived in Pierre in April after a long one hundred and ten days. Hundreds of sacks of mail had accumulated, completely demoralizing the mail service.

The winter of 1881 provided an encore wth sub-zero temperatures and severe blizzards. One man said that at the Spring Valley Ranch north of Rapid City all they had to eat was baking powder bread and rabbit tracks for meat. The men were so thin, went on our informant, that a light shone through them. Another man said he saw a coyote on the roof of one of the stations. Eventually summer arrived, the cruel winter forgotten and the roads were once again busy with commerce . . . although it wasn't a good summer for freighting because it rained a lot . . . one time continuously for four weeks. The Pierre route had the muddiest trails they had ever had.

Sweltering—and wet summers, or frigid winters didn't matter to the transportation people—the wheels were kept rolling. With the advent of the railroad, the drivers, stock tenders, bullwhackers and messengers and other personnel had to look for new jobs. Many went to work on ranches, some drove drays and local stages, some became lawmen, ranchers, businessmen or started mining. Gene Decker became a newspaperman and Dick Bullock became a guard for the bullion shipments on the Homestake Railroad. "Deadwood Dick"—Richard Clark—became a local "celebrity" as did Potato Creek Johnny in a later day.

The coming of civilization didn't set well with all the men—they did not like the restraints imposed by law—not after they had been the ones to dispense justice many a time. One was Boone May who went to South America to do some prospecting with a friend and got in a fight with a man and killed him. Boone was afraid that he would not get a fair trail and took to his heels. The friend struck it rich but Boone could not return to share the bonanza.

In 1889 Billy Sample, the "last remaining relic of the bullion guard" left for Mexico . . . one by one the staging men scattered and on the scene appeared a new breed of men in shadow of the Iron Horse . . . the railroad men.

# 11
# FREIGHTING ON THE FORT PIERRE ROUTE

In 1876 there were about two thousand people in Deadwood and five thousand in the area—with all businesses represented in the various camps. These businesses needed merchandise and supplies, most of it coming from Cheyenne, Sidney and Bismarck, but it wasn't long before Fort Pierre, the shortest route, took over the bulk of freighting. The trail was closed for a brief time in June of 1876 by President Grant but efforts were made immediately to modify the order and it was opened on the 22nd . . . and the road was busy once again with oxen, mule and horse teams, the wagons heavy with freight for the gold camps. In a few months it was noted by the newspapers that for thirty miles out from Fort Pierre the road was lined solid with freight wagons and at the Deadwood end the streets were packed with wagons, bellowing bulls, kicking mules and whackers.

Merchandise would come by railroad to Sioux City and Yankton where it was transferred to Missouri River boats including steamers of the Peck and Coulson lines, to Fort Pierre where it would be transferred to wagons. Occasionally the freight was unloaded at Yankton or Chamberlain but these cities made for longer distances to travel and the Chamberlain route traversed hostile Indian country, hence Fort Pierre was the most popular transshipping point. It was noted that a train belonging to Fred Evans came from Chamberlain in 1882 (Evans also shipped from Fort Pierre), with a load of skips and machinery for the Barstow Mining Company, several hundred tons of coal for the Homestake and eight hundred kegs of powder for a Deadwood party. The trip took fifteen days. Some of the traffic that arrived in Chamberlain by rail or by wagon would travel by road to Fort Pierre and some of the individuals would opt to take a steamship up-river.

The Hills people were always happy to see the freighters come in—the more merchandise, the lower the prices. In the summer of 1876 they were especially happy to have Richard Dunn arrive with a train of nine wagons—not only because the merchandise was welcome, but it had been rumored that Dick was dead and buried and here he was in person, a grin from ear to ear and with his top-knot intact. It was not an infrequent occurrence to have persons duly killed off by the press and then show up very much alive and perhaps with a tale or two of a close call they had experienced on the trail.

Three to four boats arrived daily in Fort Pierre and soon the warehouses were filled to bulging. In 1880 it was noted that there were forty acres of freight on the prairies, some of it uncovered and unprotected. There were a number of small outfits and three big ones engaged in freighting to the Black Hills. Bramble and Miner were wholesale grocers and millers of Yankton who did their shipping via the Merchants Transportation and Freight Company, there was Hornick and Evans that became the Evans Transportation Company when Evans bought out his partner and the third was John Daughterty of the Sidney Line who later combined with Evans. In late 1880 the Northwestern Express, Stage and Transportation Company transferred their operations from Bismarck and became the fourth big transportation and freighting company, and with Evans crowded out Bramble and Miner. The Merchants Line then tried the Medora, Dickinson and Belfield routes, none of them successful and in 1883 the company decided to go to Idaho. The Waterloo of the other companies was the Iron Horse that brought in a more efficient, cheaper and faster means of transportation—which in turn was replaced with the trucking industry.

In the fall of 1859 Downer T. Bramble had opened a store on the Missouri River bank at the intersection of Walnut Street and the levee in the frontier town of Yankton. William Miner purchased an interest in the store in 1869 and the firm of Bramble and Miner was launched to become the largest merchantile house in the territory. In 1872 in company with William Bordeno, the firm built the Excelsior Flour Mill and Elevator, the largest in Dakota. When Mr. Bordeno expired, Frank L. Van Tassel joined the firm.

In the '70's when the Black Hills became one big market, Bramble, Miner and Company took advantage of it and started shipping enormous amounts of freight, first to Custer and then to Deadwood. Weekly trains left each Monday after the arrival of the regular steamer on their thirty day round trip to Deadwood—and sometimes longer if the roads were in bad condition. In 1879 the company had 200,000 pounds of freight on the road between Fort Pierre and Deadwood, in May of the same year they had over one hundred tons on the road, hauled by twenty-six wagons with trains. By October of 1880 they had hauled over five million pounds for that year and in 1883 over 26,000,000 pounds of freight. It was estimated that they had about three-fourths of the cattle and mules engaged in Black Hills freighting. The business kept growing so much that an office was opened in Deadwood with Mr. Bramble in charge. He was described as a true-blue Democrat and no fly-weight physically—in fact there were people who called him downright obese.

The Merchants Company consolidated with Jewett, Dickinson and Company, a big name on the Sidney route, and did well for a number of years.

John Daughterty did freighting from Sidney, Yankton and in 1880 transferred to the Fort Pierre route. When his outfit approached that city it made an enormous clatter with the oxen bellowing, the cracks of whips, men swearing and the noise of the wagon wheels—so much racket that the Pierre citizens became highly alarmed. They were sure there was an Indian attack in the offing—perhaps they were extra skittish because they had had a genuine Indian scare a short time previously.

The Daughterty Line became known as the shortest and quickest thorough route from the Missouri via the Chicago, Milwaukee and St. Paul Railroad to Running Water and then the Coulson line of steamers to Fort Pierre. By October of 1880 the line had transported two million pounds of freight from Fort Pierre plus a goodly amount previously hauled from Sidney. In 1881 John Daughterty and Fred Evans consolidated their freight lines and the newspaper observed that the two had the "best bulls in the best condition" of anyone else in the freighting business.

Perhaps the most colorful man in the freighting game was Frederick Taft Evans a gentleman as portly as D. T. Bramble, perhaps more so because he was said to weigh as much as one of his steers. Evans was a sharp businessman and became the richest man in the Hills said to be worth $100,000.00. He sold his $20,000 home in Sioux City and moved to Deadwood and when the freighting business declined he moved to Hot Springs where he constructed the Evans Hotel and was active in promoting the town as a health spa—he also had mining interests in the southern Hills area.

In 1884 Evans was interviewed by the local press and he told about his first attempts to start to freight to the Hills. In May of 1875 he agreed to transport the Gordon outfit to the Black Hills by the Sioux City and Black Hills Transportation Company of which he was the superintendent. When the train reached the Running Water at Garden City—the later site of Gordon, Nebraska—it was intercepted by the army and the wagons were burned. Again Evans started out, this time from Dubuque with "live freight"—seventy-five individuals who were anxious to come to the gold camps but again the army caught up with the train and members of the party were taken to Fort Randall. In 1876 a train in charge of John R. Gordon took a similar route as planned before to Custer and then came on to Deadwood where the wagons were lowered into the gulch by rope. Gordon then went around Poorman Gulch to Lead and cut the first wagon road to the mile-high city. When the wagons returned to Fort Pierre Evans met them there with a load of freight from a steamboat and in September of 1876 came to the Hills himself—he was an accomplished bull-whacker.

In later years Evans told of taking the nuts from the wagon wheels of the troops when he was intercepted—but this didn't help him in making a getaway. He also chuckled when telling how he missed out on the

Homestake pie. It was when he met two prospectors by the name of Manuel who said they had discovered a rich body of ore but needed a mill. They asked the portly freighter if he would consider bringing in one for them and they would give him a share in their enterprise. Evans hemmed and hawed, no doubt he also scratched his head and tweaked his nose; he wasn't about to buy a pig in a poke. One of the brothers hastened to say that they would be happy to give him some samples to test and he could see for himself that the ore was bonanza grade. Evans thought this was fair enough and took the samples, but for some reason or the other they tested low and he forgot the proposition. However, in later years when the Homestake was flourishing he thought about it frequently and wondered why the samples had tested so low.

Evans on his return organized his own transportation company in partnership with a Mr. Hornick and in the spring of 1878 the two started operating on the Fort Pierre route. Evans had tried the Sidney route but found the eastern trail more practical—certainly it was much shorter, especially after April of 1877 when a cutoff was made directly to Crook City that saved thirty to fifty miles. The partnership was dissolved in late 1878 when Evans bought out Mr. Hornick.

In 1878 the Evans Transportation Company outfits made six round trips to the Hills—at the time they had the capability of moving four hundred tons in one trip. Soon the company had four warehouses in Pierre and one in Rapid City and Evans and a Mr. Doud were thinking of building a steam ferry across the Missouri. A mule train was placed on the road for fast freighting and for taking second-class passengers—in August of 1879 Evans sold fifteen tickets in a few days. The oxen were used for heavy freighting and in 1879 they hauled 5,500 tons of freight. Business was so brisk that it was necessary to add thirty-five Schuttler broad-gauge wagons to the fleet and two hundred more cattle—and still the business kept expanding. By October of 1880 Evans had already delivered 7,084,000 pounds of freight and many times had outfits unloading every day of the week in Deadwood. The freight items included four large iron evaporators for the Black Hills Salt Works, twenty tons of mining machinery for the Savage Company and carloads of bottled beer. Evans and the Merchants Line also contracted for the government including flour and other supplies for the agencies.

Evans, as well as the other freighters, would solicit freight from the various railroads and then bill agents of these lines to Lead City, Deadwood and other towns. The freight would be forwarded by the railroads to Missouri River boats of the particular line to which it was consigned and then placed on bull trains at Fort Pierre—or Yankton and Sioux City and later Chamberlain. The freighting companies paid the advanced freight charges and then collected from their customers at destination.

There were small freight companies, independent freighters and "farmer" freighters—the latter bringing in hay, vegetables or grain. Some freighters would make only one trip and then settle in the Hills, others would freight for the big companies for a time and then go in business for themselves. V. P. Shoun, a six foot four inch "you'all" Southerner came to the Hills in the winter of 1875 and started a sawmill in Sawpit Gulch near the present day Central City. He had a whip saw and with a hired man constructed a mill and sold lumber for sluice boxes at one hundred and fifty dollars per thousand feet. He later became a stock grower on the Deadwood Trail, the owner of the famous Shoun Ranch.

N. L. Witcher made $2,500 on one of the first bull trains to come to the Hills. The Witcher Company did a large amount of freighting to the Hills for about five years, shipping freight from Sioux City by boat to Yankton and then to Fort Pierre. The company tried the Chamberlain route for a time—with a five pound brass cannon for protection—but when Indians took sixteen oxen, the outfit transferred to the Fort Pierre route.

Dunn and Newbanks did freighting from Fort Pierre—they had a fairly large outfit. Joe Perrelle was known as the fastest freighter, sometimes making a round trip in eight days. There was Frank McCarthy who got a little mixed up—maybe the hard trips finally got to him and he started throwing freight from his wagons out on the prairie. And finally there was Arkansas Joe, an amiable mule skinner who had a large train and there were many others—hard-working men all of whom deserve a medal for their part in providing the frontiers with supplies to the home owner, to the storekeeper, the professional man and the mine operator.

In 1880 Northwestern Transportation transferred their operations to the Fort Pierre route, and their stage business did just fine, but they ran into a bit of trouble regarding their freighting. Evans had made contracts with the Hills merchants at the rates of the previous season and Northwestern said they absolutely could not compete at these prices. They instigated a freighter's union whereby no goods would be carried at less than $2.50 per hundred pounds to Deadwood and $2.75 to Central City. Evans tried to get the union freighters to haul for him but they said nothing doing. Finally the rotund tycoon came up with a solution—he asked the military at the new Fort Sully to force the freighters to load his wagons and if they refused to arrest them. What the military did was to order all idle men on the reservation to help with the one million pounds of freight that had accumulated at Fort Pierre and they would protect them from the union people. The union people took a dim view of this and took lynch pins from the wagons and nuts from the wheels. There was much bitterness and several altercations and some of the men were evidently incarcerated. The union hired Colonel Parker of Deadwood to represent them and he was able to get them off for alleged crimes from "high treason to petty larceny."

The freight muddle in Pierre was horrendous that spring and mountains and mountains of accumulated freight built up. It seemed to rain buckets every third day which didn't help the freight laying out on the prairie. And then to top it all, the steamer Rucker arrived in Fort Pierre with five hundred tons of freight consigned to Evans.

In June Evans reached a compromise with the freighters and the wagon wheels started rolling for the Hills full strength. Traffic jams were frequent in Deadwood and at times it was necessary to call out the police to straighten them out. People were hopeful that a fire wouldn't break out at the height of the freighting season because many times the streets were so crowded that it was an impossibility to get out the fire equipment. Frequently the streets were blocked for ten or more hours. The newspaper commented that there was a solid line of wagons from Deadwood to Centennial Prairie—a mite short on the Centennial end—and this was but one wagon train of the Dunn and Newbanks outfit.

Northwestern Railroad made arrangements to have Northwestern Transportation do all their hauling and this didn't set too well with Evans who started shipping from Chamberlain, the terminus of the Chicago, Milwaukee and St. Paul. The Brules gave him a hard time as they did others; they killed ten head of his stock plus charging him one dollar per head for pasturage . . . so Evans went back for a time to the Fort Pierre route. In 1883 he made arrangements to have freight come from Chicago by the Chicago, Milwaukee and St. Paul to Mitchell and from there on the Jim Valley road to Woolsey, then to Huron where it would be placed on the Chicago, Northwestern Railroad to Pierre. At this time Northwestern and Evans charged the same—$2.75 from Chicago per hundred and $2.50 from St. Paul.

The freighting companies flourished because they had bull whackers who did their jobs—and did them well. Usually a bull unit consisted of twenty head of oxen, yoked into ten teams before three wagons—one fastened behind the other. The wagons were big and heavy, with wide rimmed wheels and high canvas covers. About twenty units composed a train . . . and the trains were often long—sometimes one to one and one-half miles. The whackers had long whips made of braided leather with wooden handles, with a flat buckskin popper at the end that would crack with the sound of a gun. The oxen had a heavy wooden yoke . . . no lines or harnesses were used. The animals were industrious and could forage for their feed unlike the mules. At night they were corraled in a large circle after they had grazed and been watered. With oxen it took about twenty-five to thirty days to make a round trip from Fort Pierre to Deadwood—with mules it would take about fifteen to twenty days for a round trip . . . some did it in considerably less time.

The wagon train was under a wagon boss with herders to assist him. Most of the outfits were portioned eight men to a mess with a cook and chuck wagon. Meals were simple, often bread, bacon and coffee, at times supplemented with wild game, beans and blackstrap molasses. The men usually slept in the wagons or on the ground, they rarely bothered with tents. In the mornings the outfit would travel about six miles, stop for a couple of hours so the oxen could graze and the men would have a meal, then complete the journey to a station or the evening stop, making an additional distance of six miles or so.

The days of wagon freighting were coming to a close in the middle '80's. The train that had been a faint whistle in the distance was getting louder each day. The era of the Black Hills railroads was approaching with the first railroad cradled in the heart of the Hills—this was the Homestake Mining Company railroad. The company never extended the lines further than Piedmont although it had a high-sounding name, the Black Hills and Fort Pierre Railroad. The main purpose of the rails was to haul cordwood, timbers and supplies to the company's operations in Lead and elsewhere and was often called the railroad that ran from a hole in the ground to a saw log. It did transport passengers for a time beginning in 1890 when connections were made with the Fremont, Elkhorn and Missouri River Valley Railroad at Piedmont.

The freighters were fully aware of the handwriting on the wall—in fact they were supplying the "rope with which to hang themselves." In 1881-2 they were bringing in rails for the Homestake that were distributed in Whitetail and Evans brought in a thirty-ton locomotive to run on those rails—one that had been manufactured in Philadelphia by the Baldwin Locomotive Works.

Almost feverishly both Evans and Northwestern kept hauling tons and tons of freight, keeping one eye on the railroad that was changing its terminus as fast as the rails could be placed. They brought in coke for Iron Hill, salt to Galena, including one shipment of 6500 pounds; they brought in skips, machinery and kegs of powder. They hauled stamps and dies, cars of kerosene oil and carloads of doors and sashes . . . the list was endless. In March of 1883, a typical month for Northwestern, they brought in 130,000 pounds and a total for the year of sixteen million pounds. In one month Evans freighted 1,500,000 pounds and eighteen million for the year. Independent freighters hauled in another thirty-five million pounds. Not only did the various companies bring in merchandise, they took out silver bullion. One month Northwestern took out twenty wagons of silver from Galena, followed by a shipment of twenty-four wagons the same month. Late in the month Jansen's bullion train got stuck in the snow when pulling out of Galena. The roads improved and Northwestern took out 75,000 pounds shortly followed by another shipment of 82,000—the latter shipment to Omaha. Most of the ore was from the Sitting Bull and when it ceased operations a good source of revenue for the freighters ended although there still was ore to be hauled from the other mines, including Bald Mountain mines that were coming to the front.

In April of 1883 Evans bowed to the inevitable and established an office in Gordon and a couple of months later moved to Chadron. Northwestern also established an office in Chadron. However, there were passengers coming from Pierre and as late as June some company, firmly believing in the advice "Get a horse!" was planning a stage from there to the Hills. The wagon road from Pierre was used until well into the 20th century but gradually farms were established, fences put up and new roads constructed. The way stops were abandoned and most of them tumbled down

. . . or were demolished.

The railroad reached Buffalo Gap where it paused for a time, then it moved to Rapid City (Archie St. Clair had the honor of hauling the first load of freight from that town in July of 1886), Sturgis and on November 3, 1888 reached Whitewood. A year earlier Evans had said, "The time has come . . ." and took his mules to a farm near Sioux City and planned to sell them in the spring. He also fattened his oxen for the market. He had often dreamed of becoming a rancher but when his freighting days were over he remained a brass collar when he moved to Hot Springs. Evans earned a secure berth in the annals of Black Hills freighting by providing safe and dependable transportation with very little loss to the shippers for a decade.

Northwestern lost their wood contract with the Homestake, then the mining company's contract for hauling bullion and in 1888 their mail contract expired . . . but they "hung in" as long as possible. Finally they, too, said, "The time has come . . ." and sold twenty-five wagons and fifty mules and moved west where they took charge of the Rock Creek Junction line. Northwestern also deserves a gold star in transportation history—the last big freighting and staging company on the Fort Pierre route.

In 1885 Homestake gave a one-year contract to Wells Fargo to transport their gold and then later extended the contract. Wells Fargo was a famous name in the west—in 1885 they operated along the Northern Pacific from Portland to St. Paul—in fact they operated over thirty-five thousand miles of railroads, steamboats and stage lines.

A fitting epitaph for the freighters was published in a local paper.

"The Iron Horse glides over the hill,
"Goodbye, my lover, goodbye,
"It physics the freighter worse than a pill,
"Goodbye, my lover, goodbye."

# 12
# ON THE TRAIL FROM FORT PIERRE
## From the Missouri to Deadwood

Travelers left Fort Pierre by a trail going through a narrow canyon and climbing upwards for approximately eleven miles to the top of a mesa where a stop was located on Willow Creek; however, some travelers would stop sooner for a breathing spell at Tex Hemphill's Road Ranch five miles out of the port city. Here Tex served excellent meals for twenty-five cents—a frequent customer to partake of those meals was said to have been Calamity Jane. The more sturdy pilgrims went directly to Willow Creek on a road of pure, unadulterated gumbo—no wonder it took one outfit seven hours to reach there from the river. In 1877 it was noted that practically every night an outfit returning from the Hills would be camped on the creek while the foreman went to Fort Pierre to make arrangements for a new shipment. Here at Willow Creek feed for the animals was more plentiful and certainly less expensive than at Fort Pierre.

In 1881 the stop was attended by a Mr. and Mrs. Norwood. One day an Indian named Pike Monk and his wife attacked a man known as Hawk—and Mrs. Norwood became so unnerved that she jumped on the Pierre Stage, which was rather unusual. Most of the frontier women could cope with a fight, with unruly and rude—and often "pickled" customers with utmost equanimity—perhaps Mrs. Norwood was a newcomer to the area.

The trail left Willow Creek over a divide to the tributaries of Lance Creek where the main channel was crossed just below the forks, nine miles from Willow Creek. This was a regular stop in the staging days. The site is marked by a sign placed there by Edith and Roy Norman, dedicated history buffs, who have placed more than fifty signs on the Fort Pierre-Deadwood Trail up to the Pennington County line. Later maps show that Meer's Road

Ranch was located here—a good stopping place. The saloon is still standing.

In 1883 a little girl named Alice Day Pratt traveled this stage road, an account of the trip is in her book, "Three Frontiers." She and her mother boarded the stage (Northwestern) in the village of Pierre to join their father and husband who had come to the Black Hills in 1879. Miss Pratt described the horses drawing the brightly painted red coach as four snow-white, perfectly matched and groomed animals that would gallop theatrically into town and leave it in the same manner, but on the road went at a rapid trot. They were perfect horses she maintained and had perfect drivers.

The little girl was intrigued with the plains of Dakota—level land covered with curling buffalo grass and liberally sprinkled with flowers including painted daisies and prickly pear cactus. There were bright-eyed loping jackrabbits and meadow larks "shouting their jubilance." Strewn over the grasslands in every direction were the bones of buffalo that had roamed the area and then had been hunted indiscriminately by the white men.

The white horses were replaced with four equally handsome chestnuts, which in turn were "spelled" with a team as "black as night." At one time toward the end of the trip the horses were bays with black tails—all beautiful animals. Miss Pratt did not list the stations by name; however, we do know the road continued to Frozenman Creek, which was crossed a short distance south of the present Hayes. The original town was on a beautiful site on the creek, which today is a fat, lazy body of water. The buildings of the original townsite, which included a post office, saloon, store and other buildings, are marked with wooden posts today.

Ten miles west of Hayes was located the Gray Ranch, here occurred several grizzly murders although only one was ever officially confirmed. A sheepman, by the name of William Kunnecke, purchased the ranch and stage station in 1900 or 1901—this was after the stages had been withdrawn by Northwestern but there still was local traffic. Kunnecke was reluctant to pay his hired men and decided it would be better to dispatch them to the Great Beyond, thereby saving a considerable sum. His downfall was a dog left behind of one of his hands and the following employee, William Rohrbecker, wondered why his predecessor would have left his four-footed friend on the ranch. Investigation disclosed blood stains and the young man filled with fear and horror took a circuitious route to Pierre and then to his home in Iowa—his employer was trying to prevent his escape. Young William's family persuaded him to share his information with the authorities and in due time Kunnecke was brought to trial and convicted. When he was serving his sentence in the Sioux Falls penitentiary he walked off one day and was never found.

Adjoining the old Gray Ranch is the modern farm of Gene Armstrong who has thousands of acres under cultivation—in fact it is believed he has more land in farming than any other private individual in the United States. (1976.)

The Plum Creek Station, five miles from the Frozenman stop was first known as Waterhole Creek and had an excellent camping site, with good water holes in the creek. A short distance away was the telegraph line that went down to Fort Bennett and Nebraska—originally the poles were about twelve feet high, today—if you can find one—they are midget-sized. The telegrapher was Harry L. Borst.

In 1882 a freighter from Chamberlain got as far as Plum Creek when he had his horses stolen which detained him for several weeks until they were recovered. He then arrived at Power's Mill and one of the horses broke a leg—and the remaining ones got pinkeye. No doubt, the freighter muttered something about, "It never rains but it pours," loaded his freight on someone else's wagon and when once again Hills-ward probably growled, "Better late than never."

The Plum Creek crossing is marked by a sign erected in 1951 by Matt A. Brown of Chamberlain.

The trail continued to Mitchell Creek, thirteen miles away on mighty rough roads even in dry weather, due to the deep ruts. In time the ruts would wear down with the heavy trains and the trail would be fine—until the next rain. In 1885 the stage company put up one hundred tons of hay here.

A road ranch in later days in the area was the Van Cleave place about three miles southeast of the little town of Ottumwa—there are no buildings left on the site today. A granddaughter of the Van Cleave's, Mrs. Bernece Holly of Rapid City, wrote, "My grandparents, Cyrus and Ina Van Cleave, came out from Blakesburg, Iowa, and since my grandmother was a wonderful cook she was encouraged to open the eating place for the benefit of travelers going from Pierre to Deadwood. I don't know how long she ran the place as she was stricken with cancer and soon passed away."

The trail went to Medicine Creek, first known as Big Cottonwood, that was a dinnering place. Northwestern had improved their services a great deal after they moved from Bismarck as learned from Alice Day Pratt's account of a dinner stop on her trip to the Hills. Whether she stopped at Medicine Creek or another station we don't know—but the meal was undoubtedly typical. She described the meal as smoking hot and very good. Delicious pork and beef, a variety of canned vegetables—since the stops were on an Indian reservation no farming or cultivation was permitted. The piles of tin cans behind the stage stations rose almost as high as the houses—a monument to the staging days. There was delicious homemade bread, butter, milk and "thick, generous pieces of dried apple pie." Topped with huge cups of black coffee. A tasty meal for hungry wayfarers.

Shortly one entered the Grindstone country, one of the best known sections of the trail. The buttes here are dotted with round stones of sandstone and resemble the grindstones used by the Indians. The region got its name from them but the village of Grindstone is said to have received its name from Grindstone Carrie, a popular waitress in a frontier eating place—who, without a doubt, got her name from the grindstones!

This country can be reached by going ten miles north of Philip on the highway and then turning west. Almost immediately can be seen three buttes to the northwest—typical of the buttes in the vicinity. Near one of these buttes was a stopping place in 1879 for a party of freighters attached to Clay's Bull Train. While taking a break—or perhaps camped for the night—the men played a game of cards on a "grindstone" table. A quarrel ensued and a Mexican by the name of Bicenti Ortez, a night herder, was shot. He evidently lived for a time as he was buried a distance to the southwest; his grave in the middle of pasture land now owned by Howard Riddell . . . the place faithfully recorded by one of Norman's signs. The man who did the shooting was James Gilmore, who was eventually hanged in Deadwood . . . or as they sometimes say, "stretched some hemp."

Going further on the old trail the ruts are still evident where the road divided—the first "divided highway"? However, the road was likely moved over when the original road became too rutted for easy travel and another was started a short distance away. From the road of today a white marker can be seen in a field (land owned by Leo Nixon) of a grave site of four freighters killed by Indians. A freighting outfit going to the Hills had a mare go lame and the animal was left here to recuperate. When the men returned the mare was still in the vicinity but instead of picking her up then and there they went further to camp and four men returned in a wagon to get the horse. The men were having a good time singing when they were ambushed by Indians and all four killed. Years later it was said a figure of a woman frequently appeared here and then melted away—a ghost mourning the long-ago killings. People in those days were highly superstitious and ghost stories abounded—some of them extremely "far-out." The creek became known as Deadman's Creek.

Another tragedy occurred on the trail when a night herder belonging to Isaac's train was run over by a wagon when he was asleep. In those days the men slept under the wagons or on the open ground—they rarely bothered with tents—so this type of accident was not surprising.

Further along the trail on Dirty Woman's Creek was a road stop now the site of the Humbert Ranch. In later years—about 1906 to 1913 there was a post office here when the place was called Climax. Harry Smith had the post office and operated a little store. Mrs. Mabel Kiel of Philip—a great-niece of Smith, says she wonders why there was a post office here as it was only three miles to the Grindstone postoffice. In 1915 Dirty Woman Creek went over its banks and Mrs. Harry Smith and two other people drowned . . . and Mr. Smith sold the place and moved away.

On Grindstone Creek is the ghost town of Grindstone but it is a johnny-come-lately. The original town was ten to fifteen miles east on land owned by Howard Riddell. However, a visit to the second Grindstone is most interesting—there are old buildings remaining, a decrepit barn with the foundations built of the local grindstone, a cellar, a well that had the power supplied by a windmill and also remaining is a stone pump house. In 1906 the place had a store, hall, lumber yard and a newspaper office—a fair-sized town. The land is now owned by Lutheran Knutson.

The Fort Pierre-Deadwood Trail continued to the west and crossed the upper forks of the Bad River. Here halfway between Grindstone Buttes and Peno Springs Ed Sanchez had a way stop where he offered good meals and clean sheets. (Some sources say he operated the Grindstone Stage Station before coming here.) Ed liked to play cards—but he had a wicked temper and this shortcoming cost him his life. It was in March of 1902 that two cowboys had a bitter argument with the Mexican during a poker game and left in anger. One of them forgot his overshoes and went back for them. Ed met him at the door with a gun and the cowboy, Bob Adams, shot him between the eyes. He then went to Pierre and gave himself up to the authorities there.

After crossing the Bad River a ridge was followed to Peno Springs situated in a beautiful and picturesque valley that was famous as the site of the oldest route of the white man going from the Missouri River to the west. Here the old, old trail turned southward and ran to the White River and on to the Colorado Rockies and to Wyoming. The valley was named for Pineau, part Indian, who lived in a dugout near one of the springs.

The old stage station is still standing—it was constructed of logs, later covered over with boards. In the old days there were several buildings at the site—bunkhouses and a wash house south of the main house. There was a long barn in the side of the hill, sheds and some corrals to the southwest. In later days there were a large machine shed, shop and a garage to the north and a windmill to the east. Water was supplied from an excellent well. A large cave was near the house that could store a year's supply of groceries.

In the early days Sam Williams lived here—later a man named Dawson. In 1910 Honora Anthony moved on the place, called the U Cross Ranch at that time, and lived there until her death in 1919. Her children lived there until 1927.

West and a bit south of Peno Springs in hilly and bumpy ground was the Lake Station—the lake a fluctuating body of water that has now disappeared and at the site is a field. In 1882 a freighter was robbed here of his eighteen mule train, all provisions taken plus two company horses, the work of some white brigands. These outlaws were suspected of being responsible for two other robberies—one at the Cheyenne Crossing and one at Mitchell Creek. The Lake Station was discontinued as a stop but the site was still known by that name in 1884 when O. P. Johnson struck ore nearby.

The Cheyenne River, a short distance to the west, was reached over hilly terrain that in wet weather made for exceedingly hard travel and it sometimes took the mail coach many days to get there from Peno Springs. The Cheyenne River bottom had an excellent gravel base but the steep banks to get down and then pull up again were most difficult and it was necessary to roughlock the wheels. Many accounts tell of the hard time individuals had crossing the river and of the delay waiting for high waters to recede. One time a wagon train was waiting at the crossing when the horses became frightened during a mighty thunderstorm and pulled up their picket pins and took to their heels. The men in the party went in pursuit. The river waters rose and encircled the wagons—and during this time Mrs. James Thybo gave birth to a son. She called for help but the storm made such a noise that the women in the other wagons did not hear her and the new arrival appeared on the scene squalling as loudly as the storm outside the wagon and without help from the neighboring women. The baby was a fine little fellow whom the parents named James. Mrs. Thybo had several more children, including a daughter, Marie, who wrote several books, including a book of poetry.

There were several ranches and several crossings including the Allen Ranch that was known as one of the best and another one nearby owned by Tom Madden—evidentally an unsavory one as the newspaper termed Tom as a keeper of a notorious ranch, accused of harboring road agents. This was looked into by the first grand jury assembled in the Hills. Tom, at one time, was a saloon keeper in Rapid City . . . perhaps after he sold his ranch.

The ranch was purchased by Zack Sutley and a man named Endicott. Zack did the buying for the store and had charge of the haying—a lot was sold to the pilgrims. Endicott did the cooking and looked after the store and ranch. Zack and a neighbor, a man named Smith, cut hay together—each furnished a hired man, one team and in one summer were able to get one hundred tons of hay put up. Zach also purchased hay from the ranchers in the area at five dollars per ton and sold it to the travelers passing through—undoubtedly with a "handling charge."

Zack told of a horse he had that would swim the Cheyenne. One day Calamity Jane (who really got around!) borrowed it in order to get across the river in a hurry . . . seems she had been selling liquor to the Indians and the marshalls were hot at her heels. Another incident Zack related was of a man buried in a secret place nearby who confessed—before his demise, of course—of drowning an individual named William Morgan who had threatened to expose the secrets of a secret faternal order. Eventually Zack sold his interest in the ranch and was placed in charge of a freighting train.

There was a telegraph station here from Fort Sully and on to the Hills with two enlisted men on duty.

R. William Smith had a ranch that became Smithville. Later Frank Cottle became "Mr. Smithville" when he had the store and post office—and what a store! Everything was here for sale—shoes, groceries, clothes, hardware, farm machinery—much of it purchased in carload quantities. Cottle became known as Carload Cottle. He claimed to be of the aristocracy and he certainly looked the part with a handsome beard and moustache. He had the habit of clearing his throat every time before he spoke—which may have irritated some individuals. Cottle sold the townsite to Harold Wilson and moved his store and post office to the present main road—the place known as Elm Springs.

Smithville is another interesting place to visit today and follow the river road where stately cottonwoods offer coolness from a hot sun. There are also some willow and ash here and with lot of fallen timber. Birds are in plentiful supply—owls, woodpeckers, chick-a-dees and killdeers. In the outlying meadows, meadow larks sing their sweet song to all who will listen. The bottomland meadows had—and have—good grass where the weary cattle could forage in early days and many outfits used it as a wintering place for their stock.

In the area are many springs, one called the Holes in the early days mentioned as being five miles west of the Cheyenne. The Moore Ranch is west of the river—here is a spring that pumps five hundred gallons of water every twenty-four hours—the water going into a large dam. Mr. Moore's daughter and son-in-law, Dorothy and Jack Gravatts, live on the place.

Twenty-five miles from the Cheyenne was the Shoun Ranch, owned by V. P. Shoun, an early-day freighter. (The "V" in his name stood for Vest—perhaps that is why he preferred to use his initials.) He worked for the Pratt and Ferris freighting outfit, worked for a time in the Hills and later had the road ranch on Elk Creek. There are a few buildings on the place—sheds and a barn. Later V. P. moved to New Underwood where he ran a general store and had the mail contract between Viewfield and New Underwood.

A great-nephew of V. P. Shoun lives in the area today—Allen Shoun who has a place north of the early Shoun Ranch. North of his home are some trail ruts—another of the interesting shortcuts or whatever—or it may have been a trail used by the early homesteaders. In long ago days there were no roads as we know them today—only trails, so tracks can be found all over the prairie lands.

The main Fort Pierre Trail went west from Shoun's Ranch and to the Tarrant Ranch, a stop in later days. This part of the trail is marked by two signs erected by the Lakeside Old Settlers organization as part of the bicentennial effort . . . one is south of Viewfield and the other is toward Wicksville.

The Tarrants came here in 1895 to establish a ranch and they also operated a way station providing food and lodging for the travelers. Mrs. Tarrant would fix a meal at any time of the day—or night, charging twenty-five cents per meal although she finally did raise the price to fifty cents. Overnight lodging was a quarter and most of the people carried their own bedding.

A short distance from the road ranch is Washta Springs on land that belonged to Fred Holcomb who came to the area in the early days and acquired considerable acreage—the site of the springs is on land belonging to Elsa Rowe of Spearfish who is a descendant of the Holcombs. Joe Forest has rented this land for thirty-five years and about fifteen years ago he said the spring had deteriorated to a mud hole. He brought in heavy equipment, did major work on it and enclosed it in a casing—and today the springs, once again, flow with good, clear water that doesn't freeze even in super cold weather. Joe said he heard of a stage station about three to four miles to the west of the springs but had never been able to find evidence of one—bits of pottery and glass, square nails, dugouts, etc., that usually mark the old stops. In 1877 Washta Springs was given as eighteen miles from the Holes, then seven miles to Box Elder, eight miles to Rapid City and forty miles to Deadwood. A later account mentions a stop known as Lone Tree, located north of the Lone Tree School and a few miles east of Box Elder.

Rapid City was an important stop on the stage trail and described by Miss Pratt as a pretty village, located in a beautiful little valley, "through which a rapid stream flowed, bordered by sizeable trees—Rapid Creek." The stage made a grand entry here—as Miss Pratt described it, "Our bugle sounded and our team galloped up the single winding street and came to stop in front of the only hotel." From here to Deadwood was considerable distance and she and her mother stayed in Rapid City for several days.

Frank Hebert in his "Forty Years Prospecting and Mining in the Black Hills of South Dakota," thought Rapid City wasn't much of a place in 1877. The American House was headquarters for the stage with John Brennan in charge.

Early accounts give many stage stops, road ranches and stations on the Rapid City-Sturgis road—the places changed frequently and then, too, the Sidney Stage came through and evidentally had their own stops. One of the most famous was the Spring Valley Ranch fourteen miles from Rapid City that was a makeshift building at first with a dirt roof and a stockade owned by George Adler, Fred Evans and George Garvey. Evans soon gave up his interest and Adler became the sole owner when he killed Garvey in an argument over a dog; however, the tragedy must have preyed on his mind as he soon sold the place to Captain C. V. Gardner.

The first building was replaced with a large log building that had a twenty-foot square kitchen. The main part of the building had two rooms about twenty-four foot square, the low ceilings and walls were whitewashed. A large fireplace offered cheery warmth to the wayfarers, there were plenty of buffalo robes available . . . and bedbugs that were most unwelcome. At one time a man was killed in a brawl, evidentally the result of too much drinking in the barroom.

When Captain Gardner purchased the station he did not sell liquor—perhaps this was due to the influence of his brother who was a preacher. The captain kept a strict eye on his daughters and did not allow them near the barn where the walls were decorated with pictures from the Police Gazette . . . the Playboy of that day.

Gardner was an up-and-coming man, a credit to his community and to the Black Hills. Miss Pratt remembered him and his wife, Mary, as hospitable and warm people. The Captain liked social gatherings and frequently lined up picnics and parties for the residents of the community, which provided a social life for the hard-working pioneers who had little time—or money—for much socializing.

There was a stage station at Big Springs near the present Highway Rest Stop that was used by the Sidney stage—and perhaps by the Fort Pierre stage in the early days. In later years there was a roadhouse nearby run by a woman named Mrs. Meeks. At one time when a survey party was working here, Mrs. Meeks took her shotgun and told them to move on. The engineer in charge thought for a time then invited her to their camp for further discussion and she accepted. When she arrived at the railroad camp perhaps they offered her a cup of coffee and Mrs. Meeks laid down her gun—which the men seized and asked her to get lost. But the lady was not intimidated—when the tracks were placed she would grease the rails—a matter of "getting even." At one time a freighter threw her in the creek when she tried to prevent him from going past her place.

Going further north was the Bulldog Ranch established in the summer of 1878 by the McFarlands. However, it was on government land and Uncle Sam told the family they would have to move—and take their home with them. They did. They carefully numbered each log and rebuilt the house seven miles east of Tilford where it still stands. It is said that Mrs. McFarland had bulldogs that she would sic on the freighters whenever they tried to abscond with her chickens. The freighters would take their long whips, wrap them around an unsuspecting hen's legs and draw it in the wagon. Mrs. McFarland became known as Madam Bulldog and the nearby creek also bears the name of Bulldog Creek.

At one time Madam Bulldog prepared a meal for a taciturn customer who must have enjoyed the repast as he left a twenty dollar gold piece in payment. The customer was one Sam Bass, a notorious outlaw.

A newspaper of the day mentioned that a Mr. and Mrs. Joe Erb ran the "old Bulldog Ranch," and in May of 1880 Mrs. Erb was settled in the new Bulldog Ranch where she served a good meal to some wayfarers. Evidentally Mrs. Erb decided that everything new was better and forwith discarded Joe—on her own motion—and married a man named Hammond. In September of 1880 Gale Hill, a deputy, was dispatched to the ranch to arrest Mrs. Hammond, nee Mrs. Erb, on a charge of bigamy. Mrs. Erb became exceedingly irate and covered Gale with a "three foot six-shooter," while his rear was protected by the "insinuating nuzzle of a Winchester, presented at him by the 'Kid' attached to the place." Gale decided to withdraw as he did not "wish to force any lady, much less Madame to accompany him against her will" and Mrs. Erb held the fort.

Madam got in trouble again and two deputies were sent there to collect seven hundred dollars she owed Fanshawe, the post trader. The men arrived at the ranch with much trepidation but were most pleasantly surprised when Madame acted as a lady. Not that she reformed and was genteel for "evermore"—this would have been too much to ask, as a couple years later she got in hot water again when she appropriated some buffalo robes from Uncle Sam.

In June of 1882 the Bulldog Ranch, located on the "forks of the Sidney and Fort Pierre roads below Fort Meade" became known as the Pleasant Valley Ranch that had accommodations for the traveling public. We don't know who the operators were at this time; however, in 1883 the owner was Don Waln when there was a fire at the ranch that did considerable damage to the buildings and destroyed supplies. The fire was traced to a defective flue.

In 1887 J. D. Hale bought the buildings of the Bulldog Ranch and moved them to a new town—Tilford, where there was a handsome railroad depot, nine sawmills in operation, a flouring mill, bank, business houses, a newspaper and a hotel being planned. An application had been made for a post office.

From the Pleasant Valley Ranch the trail went toward Sturgis and to Crook City in the early days and later from Sturgis through Boulder Canyon.

Freighters who had merchandise for Rapid City followed the stage road to that town and the freighters going to the Northern Hills used various trails and cutoffs with the main trail following Elk Creek to the west and joining the north-bound stage road north of Spring Valley. Some took a shortcut road about a dozen miles east of Spring Valley and then northwesterly to Fort Meade. Another shortcut road used by the freighters turned northwesterly about eight miles from Viewfield—this road apparently used in the early 1880's. A still later road and used extensively in homestead days was the Alkali Trail—in fact there were two Alkali Trails—a lower and an upper one that followed Alkali Creek to Sturgis.

John McFarland tells of his grandparents McFarland crossing the Cheyenne River about four or five miles upstream from the present site of Wasta and then taking the Fort Pierre-Deadwood Cutoff that followed the Box Elder Divide. They camped north of the Melchert Ranch at the lakebed—this was in July of 1878.

There was another cutoff known as the Witcher Cutoff that was used after the Fort Pierre-Deadwood Cutoff was abandoned—it cut across Section 7, 3 W, 8 East, and cut across Elk Creek a tiny bit north of west to a point north of Piedmont.

It is easier to follow the stage roads and to learn where the stations were located as they usually stopped at designated places for meals, changing horses and for watering stops and only stopped at other ranches in emergencies. The freighters moved slowly, they stopped frequently and used stopping places that were convenient—and usually they made their own camps on the prairies. Some freighters made arrangements for hay stops, others purchased hay along the way and many relied on the grass that grew along the route. The oxen were patient, hard-working animals and could get along on grass; the horses and mules needed hay and grain to keep them going.

Going back to the main freight trail, it followed Elk Creek west and a shade north, keeping on the south side of the creek. In some places it can still be followed, on land that has not been under cultivation. An excellent place to see the trail is about a dozen miles north of Rapid City where a giant marker at the Four Winds Crossroads on the Elk Creek Road gives the names of the local residents—the Harringtons, Kammerers and other area ranchers. The old ruts are clearly visible south of the present road, they then cross the highway and go toward the north to the present Les Sherman Ranch while the present highway—as did the early main freight trail—continues to the west, past a small body of water (Seavey Lake) that in previous years was much more impressive and even now becomes a fairly respectable pond in the wet spring months.

The Sherman Ranch was the site of a road ranch, a log building with a sod roof, operated by E. D. Smith. Later the Burtons lived here and still later the Hales. Mr. and Mrs. Sherman came in 1926—at that time the sod roof had been replaced and the outside logs covered with wide boards. The Shermans stuccoed the outside and finished the inside; however, the old logs can be seen in the ceiling and on the outside overhang. The house has no foundation, and now, about one hundred years later the logs are beginning to settle a bit! Today there are no trail ruts on the place but many years ago the Shermans uncovered an old rip-rap bridge (one built of logs) on the ranch.

Long ago a group of freighters were here at the road ranch playing a few hands of poker—well, maybe more than a few hands because they drank up all the spirits on the place. A certain player still had a powerful thirst and talked one of the men into going to Rapid City for a bottle of whiskey to quench that all-consuming dryness. This player was none other than Calamity Jane who was "one of the boys" when playing cards, freighting or doing a bit of elbow bending. But in times of illnesses Jane could be a kind-hearted and compassionate woman, faithfully nursing the sick as she proved in early-day Deadwood.

Somewhere on the ranch are two graves—one is of Smith's wife and the other of his baby. There are no markers showing the location—Smith claimed he knew where the graves were and it was no one else's business!

Another stop for the freighters was the present George Blair Ranch, with the early owners being the Cases and Mattsons—the Blair family purchased the place in 1907. The home is one of the oldest buildings standing in Pleasant Valley. The ranch was located on the Rapid City-Sturgis stage trail that went just west of the house—present Interstate 90 goes considerably to the west.

The kitchen part of the Blair home was a part of an old roadhouse that was in business in the early '80's. The yard in front of the home sports a sunken spot—this was near the watering trough where the freighters and travelers watered their animals. One of the interesting things to see is an iron plank grown fast between two giant trees a short distance east of the ranch home—the plank used in shoeing the oxen—there evidentally was a blacksmith shop closeby. To the north was a greenhouse where vegetables were grown to sell at Fort Meade and at Deadwood and other mining camps where there was a great demand for them.

East of the Blairs was an early trail to Fort Meade—one of the shortcut freight roads, in fact one can still get to the fort by this road. There was another shortcut road to Fort Meade closer to Sturgis, as shown in early-day maps.

The shortcut freight trail that was already mentioned as branching off the main trail west of Viewfield can still be followed by the ruts—for instance on the Edgar Harrington Ranch twenty miles southeast of Bear Butte. This was high ridge country that the freighters preferred in the winter time because of less snow but in the summer they used the lowland routes to be near a source of water. We know this is high country because near here is the highest point above sea level east of the Black Hills, the place now marked with a cement and iron marker placed there by a United States geological surveying party who did their surveying by lighting beacon fires on the various peaks.

The trail angled to the northwest, went past the Eveleth Horse Ranch, toward Alkali Creek and then turned almost west until it reached the freight trail and then went north to Fort Meade. This trail wasn't used as much as the others apparently because the ruts are not as deep as on some of the other trails.

The Alkali Trail crossed Alkali Creek shortly before it joined the Belle Fourche. At the crossing lived the McPhersons—thirty five miles east of Sturgis. The Fort Pierre-Deadwood Trail passed between their house and barn. When the family moved here in 1890 there was a frame house on the place and someone had started to build a barn which they finished with lumber hauled in from the Black Hills. The traffic consisted of people going to the Hills and in homestead days it was honyockers.

A short distance from the McPhersons the road split into an upper and lower trail . . . the upper one followed the ridge between the Belle Fourche River and Alkali Creek for several miles—this road used in wet weather. The lower road followed the creek more closely, passed Sulphur Springs and also went to Sturgis.

An early-day resident on the lower road was Domonic Keffler who came from Mt. Carmel, Iowa and built a home here for his family, a large family—there were fifteen children. One of the sons later had a ranch on the Upper Road.

Fifteen children or not, the family served meals to travelers—at first at no cost but in later days they asked for twenty-five cents per meal. Indians used to stop here frequently and show an order from the quartermaster at Fort Meade stating that Mr. Keffler was to feed them—which was fine except the order was a fake. The matter was eventually straightened out and Mr. Keffler was told he did not have to feed the red men—that they had plenty of provisions to see them to their destination.

The water here has a high sulphur content which is good for cattle—there is usually also plenty of feed in the pastures. The land is good for raising crops—that is, it produces abundantly with good farming practices. Which is much easier today with modern equipment—quite a contrast to Dominic who plowed his fields with oxen.

This covers most of the main trails, but there were others—shortcuts used by the freighters and travelers, mute evidence of them remaining in grass-covered ruts that criss-cross the prairie lands of Dakota.

The July 10, 1878 route of the Fort Pierre Trail was given by the Black Hills Weekly Journal as: Fort Pierre, Willow Creek, Lance Creek, Plum

Creek, Medicine Creek, Bad River, Lake Station, Cheyenne Crossing, Lone Tree, Box Elder, Rapid City, Spring Valley Ranch, Sturgis City and Deadwood.

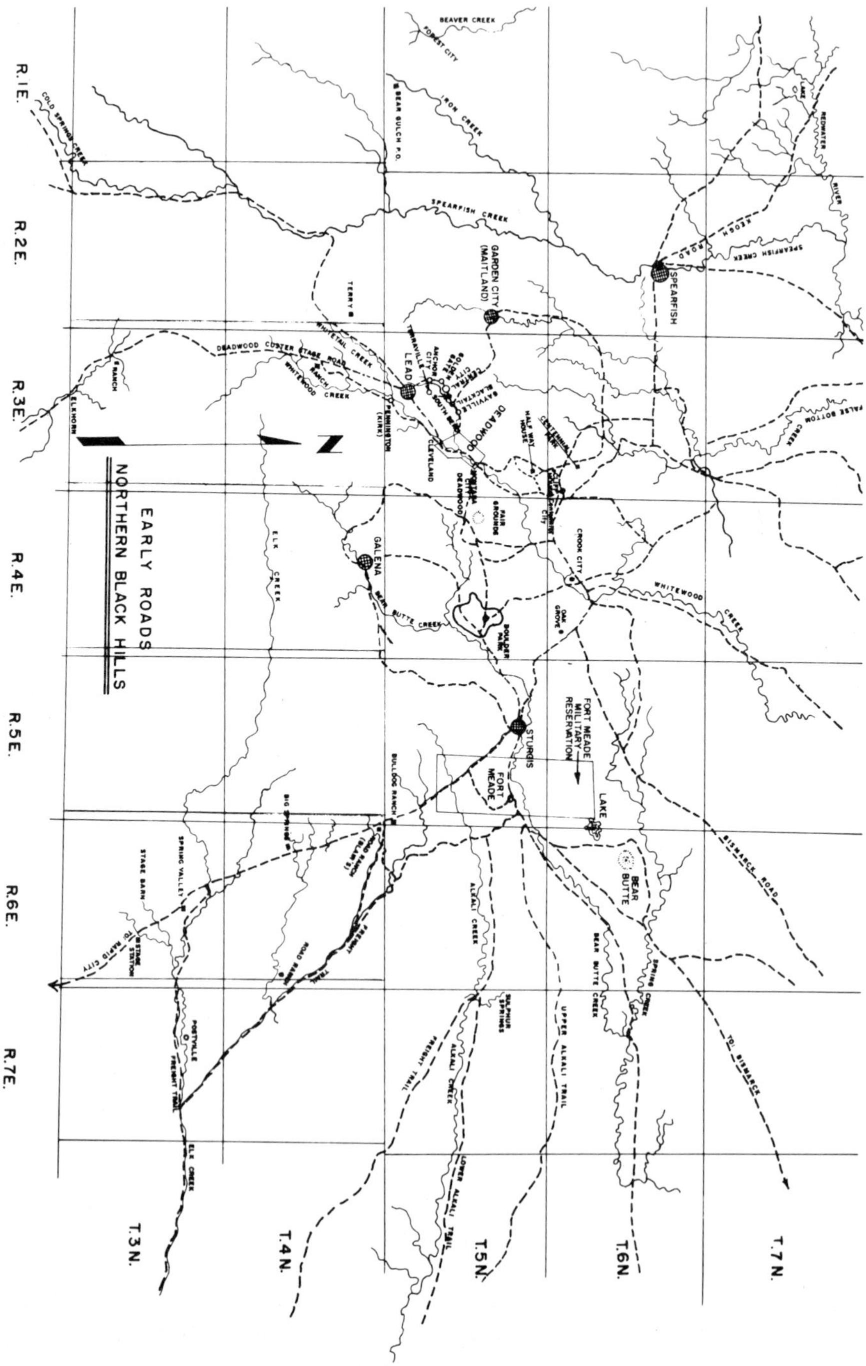

*Lonesome country on the Fort Pierre Trail. In the early days there were farms dotting the entire area—now only a few ranches remain.*

*Slips, similar to the one pictured, drawn by horse teams, helped build the roads in the west. This one is perched above the Cheyenne River.*

*This photo of an old school tells a story of the open prairie country on the Fort Pierre Trail. This is near the Edgar Harrington Ranch.*

*Ferry on the Missouri . . . the first "portable" bridges. The river was a highway to the Montana country as well as points in between—the first steamer ascended the river in 1830. Fort Pierre was on the west side of the river, Pierre on the east that was laid out as a townsite in 1880 when the Chicago Northwestern rails were approaching at the rate of one and one-fourth miles per day—Photo courtesy: Deadwood Centennial Collection.*

*Fred Evans who had one of the large freighting outfits coming to Deadwood. He did most of his freighting from Fort Pierre but also tried the Chamberlain Route but the Brules discouraged him using this trail.*

*The site of the early town of Hayes on Frozenman Creek where there was a crossing in the early days. The locations of the early buildings are marked with signs similar to the one in the photo. When the modern highway was built to the north the town moved there.*

*Kunnecke was a sheepman who did away with his hired men instead of paying them their wages. This was the original site of the Gray Ranch that Kunnecke purchased. The sign was erected by the efforts of Mr. and Mrs. Roy Norman who are pictured here.*

*Plum Creek was an important stop in the early days . . . the creek sometimes known as Waterhole Creek. This marker was erected in 1951 by Matt A. Brown of Chamberlain.*

*The old stage, freight and Indian trails still crease the prairies as do these ruts near Hayes. Surprisingly enough sometimes the ruts can even be followed on plowed ground.*

*Bicenti Ortez was killed a short distance from here when playing cards with men in his freighting outfit. Mr. and Mrs. Walter Sandal are looking at the grave that is located on pasture land belonging to Howard Riddell.*

*This is an old barn at the second Grindstone—the first town was located some miles to the east. The foundations of this building are of stones resembling the grindstones of the Indians.*

*The old stage station at Peno Springs—the building was of logs covered with white boards. In the early days the building was used by the U Cross outfit.—Photo courtesy Mrs. Lucylle Bloom.*

*The Cheyenne River—an extremely hard river to cross due to the steep banks. Several places were used to make crossings.*

*Here along the Cheyenne River tall cottonwoods offer welcome shade to travelers or did when the road went through here. When the modern road was built to the south Smithville faded away.*

*Smithville was named for R. William Smith who had a ranch here. In 1882 elections were held in the town and at Shouns Ranch to the west—the elections should have been held in only one place.*

*Five miles from the Cheyenne were springs called the Holes in the early days. This is the Moore Ranch that has a mighty fine spring that provides five hundred gallons of water every twenty-four hours.*

*The Shoun Ranch site—a good stopping place for wayfarers in the early days . . . and in 1882 it was noted as a good place for hunting with plenty of ducks, chicken and rabbits in the area. In the photo we have Mr. and Mrs. Allen Shoun who live a short distance away—Allen is a great-nephew of the original owner of the ranch.*

*V. P. Shoun who established the ranch. He did freighting in the early days and for a time worked in the Hills.*

*Washta Springs that has clear, cool water can be seen in center. Joe Forest who has rented the land for thirty-five years said he had heard of a stage station three to four miles to the west but has never been able to find a trace of it.*

*Edgar Harrington is standing in the ruts of one of the trails to the Black Hills that was used around 1884. From here Bear Butte can be seen twenty miles to the northwest. Some say that early travelers were told to "bear toward the butte" hence the name of that famous landmark.*

*Here is Bear Butte sporting a little cloud at the top.*

*Mr. and Mrs. Sherman in front of their home on Elk Creek that was a stopping place for freighters and travelers in early days. The original logs can be seen in the overhang. One of the freighters who stopped here long ago was Calamity Jane.*

*This was the site of the original Bulldog Ranch operated by Mrs. McFarland. It was near the National Cemetery and the modern highway goes over the site now.—Photo courtesy: Mr. and Mrs. John McFarland.*

*The Blair Ranch that in the early days was a road ranch. It is one of the oldest standing buildings in Pleasant Valley.*

*Between these giant trees is a bench that was used in shoeing oxen.*

*The old road in Boulder Canyon went through Oak Ridge Cemetery, near the trees on the right. One of the many drivers who rode the stages through here was Red (E. R.) Raymond who drove on all lines from 1876 until the stages were withdrawn.*

*The Deadwood coach on the way to Pierre with the Supreme Court Judges on their last trip to Deadwood. Also on board was the first legislative delegation for the new South Dakota.—Photo courtesy: Alumni Association, South Dakota School of Mines.*

# 13
# THE BISMARCK TRAIL

Ben Ash was Mr. Bismarck Trail—in December of 1875 he and his stalwart companions, S. C. Dodge, Stimmy Stimson, Ed Donahue and Russ Marsh established this route from Bismarck on the Northern Pacific Railroad to the Black Hills of Dakota Territory. People were clamoring to get to the Hills with all its riches and Mr. Ash volunteered to locate a road. He and Mr. Dodge set out on the 19th of December and that evening were joined by the other three—the names of the party are on a marker about one mile from Cedar Canyon where the Black Hills first glimmered in the distance. (The trail itself is about two miles east of the marker.) The trail blazers arrived at this spot on Christmas Day and four days later hit Custer's old trail on the Belle Fourche River north of Bear Butte. They followed it part of the way; however, it was a circuitous route and the men wanted a more direct way to Custer City. Gold had been discovered in Deadwood and Whitewood Gulches but it wasn't general knowledge at the time, hence the road to Custer City.

It wasn't long before the Bismarck Trail was darkened with commerce—horse, oxen and mule wagons with trains, all carrying freight to the gold camps. In February of 1876, H. Ross, one of the discoverers of gold on French Creek, formed an expedition in Bismarck to come to the Hills, carrying a steam sawmill, blacksmith shop, a shingle mill, food such as flour, sugar, coffee and salt—all the staples needed by the miners. Some "mobile" freight-cows accompanied the expedition. Soon there were other outfits bringing in supplies, including one of Miller and McPherson with a train of twenty-three wagons that arrived in Deadwood City in October of 1876, as the road went there by then. Later outfits were Joe Pennell with seventy-five teams, George People with twenty-seven and Northwestern with three hundred. Other outfits had about fifty teams engaged in freighting on the Bismarck Trail.

The goods that the later outfits brought in were more "exotic" than the earlier merchandise and included safes, some weighing as much as three thousand pounds, crockery (fifty-five days from Liverpool), four thousand pounds of peanuts. The "goobers" no doubt, were to go with the huge amounts of budge dispensed in the many saloons in the Hills. Mine machinery, including complete mills, were freighted in with much of it labeled for the Homestake Mining Company that also received shipments from Cheyenne and Sidney. At one time it took only eleven and one-half days to bring a twenty-stamp mill to Grizzly Gulch; the usual time with good roads was 14 to 16 days. An incubator was brought in for an enterprising chicken farmer on False Bottom. Also arriving were brick machines, fresh butter, a hook and ladder truck and Dr. Dickinson received a fine microscope with as much "machinery as a Corliss Engine."

Practically everything that was available in the states was also available in Deadwood City; however, the transportation cost often was as much as the original price of the product. With larger and larger amounts of merchandise, the prices fell accordingly and soon compared favorably with those in the states. In January of 1878 it was noted that Northwestern had forty wagons at Crook City with over one million pounds of freight and with two hundred wagons still on the road. Even safes soon glutted the market and a number of them stood in the Northwestern warehouse with no one to claim them because at that time there was a shortage of money—either to buy them or place in them for safekeeping.

The Northwestern mentioned was the Northwestern Express, Stage and Transportation Company, the big outfit in the transportation business. The company was organized in the spring of 1877 with R. Blakely as president and C. W. Richardson as treasurer and provided passenger, freight and express service with its fleet of twenty-eight passenger coaches, one hundred and seventy-five horses, twenty-five drivers and eight messengers. In the freight department they had two hundred and fifty teamsters, six hundred horses and mules and three hundred and forty-eight oxen. The monthly payroll was an impressive $12,000—at least impressive in those days. The company constantly upgraded the equipment by adding new coaches and rolling stock and in January of 1878 it was noted that they purchased one hundred Studebaker wagons with trailers. The following month Peter Seims came in with a new treasure coach—one guaranteed to discourage the knights-of-the-road. Later in the year elegant Concord coaches were added that made for easier riding than the older styles; they were nine hundred pounds lighter and this also made for better time on the runs.

It took passengers about seventy-six hours to come from Chicago on the Bismarck line, including forty-five hours from the railroad to Deadwood. It was hoped to cut down the time to thirty-six hours, pushing day and night, but forty hours appears to have been the best time made. At first service was provided three times a week, then five times and finally daily service. When needed extra coaches were pressed into service and when passengers dropped in number, service was curtailed to five days a week . . . in the fall of 1878 the stages were coming tri-weekly. The greatest number of pilgrims came in during 1878 and 1879 when at one time twenty-three individuals arrived, followed by fifty-four—this in a period of forty-eight hours. Of the fifty-four, twenty-three of them were well-heeled Frenchmen directly from France. In September of 1878 an average of six passengers a day arrived on the northern route and the hearts of the lonely gulch men were exceedingly gladdened when a number of them were of the "gentle sex." A great many old bachelors—and some not so old—would just happen to drift to the stage depot and check out the new arrivals, and if they were young, pretty and female, so much the better. On Wednesday, October 19, 1877 one of the interesting individuals to arrive was Haynes, a photographer for the railroad who came in to photograph interesting sites in the area.

Of course there were people leaving the Hills, too; one time Lola and her troupe of two, Jeanett and Pearly, left Deadwood and arrived safe and sound in Bismarck where they were much admired for their lady-like behavior. Mentally ill patients were transported to St. Peter in Minnesota, this was before the State Hospital, called the Insane Asylum in the early days, was constructed in Yankton. The government also chartered stages to transport prisoners from Fort Meade but one time when Lieutenant Mann went to the fort to collect his prisoners, found that the birds had flown. The newspapers observed that only one man in three registered his proper name when buying outbound tickets; privacy in those days was highly prized and only the most obtuse would ask another's name.

Northwestern was an up-and-coming outfit due to good management and to employees who saw to the comfort of the passengers. Buffalo robes were provided in cold weather and an ice chest with ice in the summer months. Milk cans were stashed in the boot to provide sweet milk for the children. Equipment was kept in top shape and the stock in good condition—it was noted that the big mules, seventeen to eighteen hands high, were fat as kittens and when on the road were stabled and blanketed the same as the stage stock. The local office in Deadwood, a three story brick building, was kept up under the eagle eye of the agent, William Selbie. In February of 1878 the stable received a remodeling job and all agreed that it was the cleanest and largest barn anywhere with whitewashed interior and compartmented for the stock and freight teams. The freight building was the first building in Deadwood to receive a coat of white paint. In the office was a ticket case, the first of its kind in the Hills, a beauty that dispensed tickets to any point in the United States and territories.

In September of 1878 the old sod roofs of the way stops were replaced and some buildings were torn down and rebuilt. Bridges were constructed over the streams and coulees and new stations established at various points. A satisfied passenger noted that the messengers and drivers were thoroughbred gentlemen, the horses the best looking, best kept and the speediest, averaging nine miles per hour. Meals at the stations were good, although there were a couple of individuals who thought otherwise, but then maybe they were exceptionally finicky eaters. The menu at one stop—similar at the others—featured sowbelly, potatoes, bread and black coffee served in generous quantities, but not too well cooked. Nonetheless, Northwestern aimed to please and did a good job of it, too.

The company not only took good care of the passengers but had the interests of their employees at heart too. The stages ran the year-round but freighting ground to a halt when winter blizzards shrieked across the frozen prairies. The freighters needed winter employment and the company provided it by getting timber contracts—frequently from the Homestake that used forty cords of wood per day in their shops and giant mills. In June of 1880 a Northwestern mule train went to Boulder Park to load up with timber that they took to Pennington (the present Kirk) for building cabins and stables for the timber workers. That year the Homestake contract called for 1,700 cords of wood that the teamsters hauled three and one-half miles over the divide, making two trips per day. This road from Pennington was a steep road and a dusty one in dry weather that took four men with two sprinklers to keep down the dust, as well as to keep it in good repair.

Soon the company constructed a blacksmith shop, a wagon shop and a roundhouse for fifty employees at the wood camp. In 1881 the transportation company again received the Homestake contract to supply wood for the entire year and the workers cut 17,000 to 18,000 cords of wood.

Selbie was the overseer for the timber operations—and interested in the mortal souls of the wood choppers, mule skinners and others. He held Sunday School classes for them (no doubt many of these rough, tough men needed it too!) and even General Gidley, the manager of the stage company, received "moral instruction" when he came to town. The General came to Deadwood with the first coach and superintended the Northwestern line as long as it was in existence, even after it changed to the Fort Pierre route. After his staging days were over he went to Montana and at his death was said to weigh three hundred pounds and it was necessary to make a special casket for him.

After Selbie's staging days were over he remained in the area and worked in Deadwood and Whitewood banks. He owned land near Whitewood, in fact the present town was built on his property.

Travelers on the Bismarck road suffered from Indian depredations as did those on the other lines. In November of 1877 a band of eleven red men and one squaw attacked the coach at Elm Creek, shortly after there was an Indian attack at Sulphur. At one time a party was corraled four days at the Belle Fourche before the Indians left. The following month a party of six whites fought the Indians at the river—the affair was a draw. In January of 1879 Billy Hicks felt especially fortunate when he arrived in Deadwood with his scalp intact as there was only one gun in the train. George Glover, of Lead, whose father did freighting in the early days, tells of the men sticking wooden "guns" from the "portholes" in the wagons and the Indians would be tricked into believing that armed men were with the train. A passenger in 1879 noted that the messenger did not carry the usual shotgun but had a Winchester rifle and the passengers had carbines.

In 1877 the Sioux were especially hostile and Major Whitehead who had established a colony on the Grand River between the Hills and Bismarck got a cannon from Fort Abraham Lincoln to provide protection. Later he decided to come to Deadwood and started out with the piece of artillery. When he and his party arrived at the Belle Fourche station (east of the present Vale and near the site of the early town of Empire) they found the agent there mighty uneasy. He had seen a number of Indian signs—-burning campfires and smoke signals—and asked Whitehead if he might leave the cannon there. Whitehead agreed, after all he wasn't far from Deadwood and anticipated that he wouldn't have trouble from the Indians. The cannon was never needed and finally showed up in Deadwood where it was used during various celebrations. Eventually the Army Quartermaster instituted inquiries regarding its disposition and learned that it had washed away in a flood. But that wasn't the end of the story—or of the cannon—because it did show up many years later in an old channel of the creek where torrents of flood waters had carried it.

The army decided to offer protection on the stage roads and General Bradley with six companies of cavalry formed a camp near Bear Butte and a small detachment was also sent to Rapid City. The camp was in existence for a brief time under the command of General Sturgis when General Sheridan came to the Hills to locate a permanent military post. The place selected was the present Fort Meade and its construction gave a lot of business to the freighters. It was estimated that 1,085,000 feet of lumber would be needed, 200,000 lath, one million shingles, 100,000 bushels of lime and one million bricks. The fort was duly constructed and today is a government hospital.

Northwestern had an admirable record as far as holdups were concerned. The passengers were asked to deposit their money before starting out and were issued a check for the amount on the agent at the other end. Meals and lodging were endorsed by messengers and deducted from this amount and even those not taking advantage of this service paid the messengers and not the station keepers. This policy paid off because in August of 1877 when the stage was robbed a few miles this side of the Moreau by five road agents, the haul came to a measly two dollars. A few days later another robbery yielded one single, solitary copper cent. The slim pickings discouraged most of the road agents and they naturally gravitated to the other lines where the spoils were richer.

Gold was transported on the Bismarck line for only a short period of time and this, of course, was fair game for the brigands. The bullion was placed in a two-compartmented safe, constructed of steel plates placed alternately with sheets of chilled iron, seven in all. The eight hundred pound safe was placed under the seat and out of the way of the passengers. A combination lock was fitted on the chest. Guards or outriders were employed, usually five of them and sometimes more, all selected for their bravery and fidelity. They used relays of ponies—light, wiry animals with a lot of stamina.

In 1877 a whopping $350,000 worth of bullion was shipped out on the Bismarck route; the largest shipment ever made on this road. Thirteen men were employed as outriders with some in the front, others at both flanks and one in the rear at three hundred yards. The treasure coach jolted over the prairies with the teams of horses going full gallop—and the treasure reached Bismarck safely. It was later learned that these precautions had been worthwhile because free-booters had planned to hold up the coach at Grand River but seeing the full complement of heavily armed men had changed their minds.

Even with all the precautions taken by Northwestern, the treasure coach was robbed two times in a short period of time and the bullion shipments went back to the southern route.

Road agents, Jones, Wall and Blackburn "worked" the Bismarck trail as they did the others, but they never molested Northwestern coaches. It was said that was because Superintendent Gidley had extended a kindness to Wall's brother by giving him a free ride on the coach when the boy had been flat broke . . . and the gang reciprocated the courtesy by leaving Northwestern alone.

The Northwestern company did not allow liquor on the route—perhaps this was due to the influence of "Deacon" Selbie. However, at wet-your-whistle joints between Deadwood and Bismarck, the men would make quick sorties to the Ruby Saloon in Fountain City, the Bank Saloon in Crook City and other watering places to sample some of their potions. One time when Eph Witcher and Howard, an outrider, stopped at the Bank Saloon, just to "keep the cold out," Howard's gun went off accidentally and did some damage. The bar owner was highly displeased and demanded fifty dollars to cover the damage. When the money was not forthcoming, the saloon man raised a scene and Eph, not a patient man by any means, got tired of the man's yapping, took out his gun and shot the bartender in the fleshy part of the arm. No doubt the victim changed his tune on "ow-ow-ow" and Eph and Howard made fast tracks from there.

Eph Witcher and his family were early freighters to the Hills—in fact Eph first came here with the Russell-Collins party in the fall of 1874 and wintered at the Gordon Stockade. We should say wintered there in December, January and part of February until he and John Gordon rode back to Sioux City to show gold that they had washed out of French Creek. The two men had promised to organize another expedition to the Hills but the army took a dim view of this, intercepted them on the trail and burned their wagons. Witcher came in the following summer and, subsequently, made many trips.

Eph was of the opinion that the Cheyenne River was navigable within sixty miles of Deadwood and floated down the Belle Fourche to the Cheyenne about forty miles below the crossing of the Fort Pierre wagon road. It was quite a trip because the party was captured by Indians and they had a hard time convincing their captors to release them . . . but they were successful.

In later years Witcher entered the grocery business in Deadwood and he also owned land in the foothills where he raised stock. During his adventurous years he lived through a myriad of experiences from climbing a tree to escape a bear to shooting scraps, including the one already mentioned.

The lack of booze on the line was treated lightheartedly by the Black Hills Daily Times in their account of Fatty Chase, Fatty Brown and a three hundred pound Frenchman who were snowed in somewhere along the road, the trio envisioned as monuments, huddled over a feeble campfire—yet smiling bravely at grief—no budge, no fun, no nothing. At that they were lucky, they survived the storm when many times others did not. Storms often caught long wagon trains by surprise as in March of 1878 when all the stock of the Pratt and Ferris train died when caught in a storm beyond the foothills. It was said the oxen died as "fleas."

Traveling in those long ago days was for the hardy—with or without a few sips of "select lightning." A description by a woman passenger enroute to Bismarck in May of 1880 is a good account of traveling in those days. "We started out in the light of the early morning," she wrote, "nine of us and a baby, in our crisp linen dusters and tidy appearance and after seventy-five miles we were on our best behavior, sitting upright with pleasant chattering and merry jokes, but at the end of the time one by one succumbed to the drowsy god until the only one alert was myself and it amply repaid me for my vigilance to see them with heads bobbing and nodding with every motion of the coach, until a sudden lurch would send them sprawling into their opposite neighbors arms." (Now that's what I call a sentence!) But on with the good lady's account. "And then in an effort to brace up and not be caught napping again was laughable.

"The eating houses were all managed without the aid of 'gentle woman' which is plainly to be seen. We are always received with open arms and a brace of dogs and still we are ungrateful enough to miss something. At one station where we thought to remove some of the dust of travel, we found a towel that one gentleman thought must be old enough to stand alone. But the most amusing part of the trip was the general look of dilapidation of all the passengers. The crisp linen was sadly rumpled while the disheveled looks and wild eyes of the pilgrim told a sad story of the results of a trip on the plains."

Our lady passenger on the Bismarck Stage took the trip with good humor but not a young tenderfoot from the east. He had been fed a lot of stories of road agents and Indians which the westerners had given him and he traveled the whole way shivering and quaking—and it didn't help when they traveled over areas where prairie fires were burning that resembled "the uttermost infernal regions." No doubt, it was with a sigh, or more likely ten sighs of relief when the coach arrived in Deadwood and the young man wasn't sporting an arrow through his hat.

The railroad arrived in Pierre in 1880 and Northwestern moved their operations to that city.

# 14
# ON THE TRAIL FROM BISMARCK
## From the Northern Pacific to the Black Hills

Bismarck was first located on the east bank of the Missouri River in 1872 but a dispute over the legal ownership of the land resulted in the town being moved a bit to the east near a United States weather bureau called Fort Hancock. Bismarck was at first called Edwinton in honor of Edwin F. Johnson and in 1873 was renamed Bismarck in honor of the German statesman, who was often called the "Iron Chancellor."

The Northern Pacific Railroad, of which Edwin F. Johnson had been an engineer, reached the town of Bismarck on June 5, 1873 and was its terminal point for half a dozen years—until 1879. In 1883 Bismarck became the territorial capitol for the present North and South Dakota and the capitol for North Dakota in 1889 when the two states were formed.

At first Bismarck had been a prairie cow town but by May of 1878 was having growing pains. The railroad had stopped at the river but in the winter lines were laid over the ice (one winter the ice was reported to be thirty-six inches thick and later forty-seven inches) and when the river was free of ice boats were used for crossing. Northwestern Express, Stage and Freight Company purchased the steamer, Denver, from George Acker to carry goods across the river . . . this was the same George who was robbed of one hundred and eighty-five dollars on the Bismarck line; however, the company reimbursed him for the whole amount. Eventually a bridge was constructed and after completion Northwestern moved their offices to the west side of the river. In 1878 several large business houses were built and handsome dwellings erected. Farming interests, too, were receiving attention and it was noted that a large colony of twenty families from Maine had located on Apple Creek Valley east of town.

Starting from Bismarck the trail followed the Missouri River bottoms for a short distance, then veered to the southwest crossing the present highway 6 in the southwest corner of Section 16, Township 137, Range 61. In the area was an alternate road—the ruts still in evidence. The first stop was made on the Little Heart with the stage station apparently located about three miles north and one mile west of the village of St. Anthony, where there are remains of dugouts.

The trail left the creek, still going towards the southwest across the Little Heart Flats toward the Oak coulee pass. On one of the coulees coming down from the south was a trench or earthworks although no one knows its history—who built it, when and for what purpose. Climbing from the flats, through the Oak Ridge pass there is a plateau where to the northwest can be seen Chimney Butte, to the southeast is Rattlesnake Hill while to the southwest are the Three Buttes and the Dog Teeth range near Raleigh—the latter named for the sharp hills resembling canine teeth.

The trail followed Chanta Peta Creek and crossed it south of Chimney Butte, it then shortly veered to the south—here is level farming land until the area of Lake Patricia is reached. Before reaching the lake several forks of the Chanta Peta were crossed—including the west fork—the Fire Heart but local residents call it Louse Creek. Dana Wright who studied the old trail, says that an early-day freighter who was camped at the crossing marveled at the size and strength of the animals that inhabited his shirt. These little creatures were companions of the trail and even if a man took a bath and burned his old clothes, bought new ones and donned them, it wasn't long before a new colony of the mites moved in and set up residence.

We don't know for certain where the trail crossed Louse Creek at Flasher nor the site of the stage station but a sign has been erected east of the town commemorating the old stage trail. The route was to the southwest, going on the east side of the Dog Teeth and crossing Dog Teeth Creek east of Raleigh, where the present Highway 31 follows the trail for several miles. The trail veered more to the west—about two miles north of St. Gertrude; it then crossed the Cannonball southwest of the present Leith, North Dakota. The river is said to have received its name from the rocks on the banks that resemble ammunition for cannon . . . rocks that are round balls from six inches to three feet in size.

Here the old trail ruts can still be seen that went down to the crossing a short distance south of the newer bridge. At the crossing, in later years was located a government stock dipping tank. The stage station was located at the present site of a picnic area on land that was owned by Jack Stewart who often mentioned of traveling on the old Black Hills Trail in the early 1900's when going to Mandan. Mr. Stewart's son, who lives nearby, states that the original stage buildings were of sod and included a big barn and two smaller buildings—one used by the stock tender and the other for supplies. Water was from a spring that is still faithfully running.

The Cannonball Crossing picnic area is enjoyed by a great many people coming here although the birds that have moved in one of the shelters take a dim view of all the human traffic. The grass is mowed and the place kept neat by the Stewarts with the state taking care of the tables, shelters and facilities.

The trail inclined to the west after the crossing and then again changed to the west, southwest to cross the divide of the Cannonball and the Cedar. In Section II, Township 131, Range 87 the modern county road follows the trail about one mile, then the old route crossed several forks of Crooked Creek and went down to the Cedar which was crossed in a corner of Grant County. There was another crossing of the river in Sioux County where there was a stone crossing. On a bluff west of the ford is a little bluff and on the south edge are the dugouts of the old station—one of the more important stops along the route. According to Rose Tidball in her book, "Taming the Plains," there were two big sod barns on the north side of the Cedar River—about five miles west of the bridge north of Watauga. There were two men permanently stationed here to take care of the horses and to do the cooking. Nothing is left of the place now except a couple of holes in the ground. Leaving the stop the crossings united and went west of an old schoolhouse where the trail can still be seen . . . it then went nearly south to the state line, leaving North Dakota where Hay Creek comes in from South Dakota.

The trail crossed Hay Creek just about a mile east of Morristown, the small town of today located on U.S. 12.

The next station was Antelope, a well-known stop in the early days. It was fifteen miles from Cedar Creek and early newspapers mentioned it being eighty miles from Deadwood, but distances weren't too accurately given in those days. Here was a large dugout barn, several houses, evidence of other structures and several graves. It is said there are some graves north of the station site and a common grave to the northwest—the burial place of several victims of an Indian raid in the early days. There is also a story of lost gold and in later years a Wells Fargo strongbox was found, although empty as a barn in a dry year. The box may or may not have been the receptacle for the gold—after the passage of so many years, we'll likely never know. We do know the stages were held up by knights-of-the-road, on the Bismarck Trail as on the other routes, because when there was gold on the road there were robbers plotting ways to "requisition" it.

There was another station on Dirt Lodge Creek about one and one-half miles from Antelope that was called the Whistle Stop Station, said to be constructed as outlying protection to the main station—and a port-in-storm for the pilgrims when they couldn't move one more foot—or wheel in the gumbo. Gumbo is the glue of the prairies, the chewing-gum wad, the molasses—the bane of travelers in wet weather. How they would have appreciated the all-weather roads of today—that we frequently take for granted.

John O'Donnell's grandfather homesteaded here in 1911 and says that there was a dugout barn with sixteen bunks at the site . . . perhaps used in days when it was headquarters for trappers. The homestead was a bit east of the station. The locations of many of the stage stations were marked by spent cartridges, square nails, wheels, oxen and horseshoes although most of the buildings were gone in homestead days. The stops had been staffed by employees of the stage company and when Northwestern moved its operations to Fort Pierre the stations were abandoned and soon tumbled down. Then cattle kings moved in the area and last of all the homesteaders in search of free land—and by then the stations were a part of past history although the trails were marked by deep ruts—many of them still evident in virgin sod.

Antelope and the Whistle Stop stations were located on dry creeks that do not have water all year; however, at Antelope there was a well. East of Whistle Stop was an interesting bridge, constructed with rocks that needed no support—a "neat feat" of early-day engineering.

The trail continued Deadwood-way and crossed Willow Creek near a spring that even now usually has water. There were trees here in the early days although very few remain as most of them have fallen victim to industrious beavers in the area. The next station was at Grand River located on the east bank. North and east of here was a large barn and nearby are remains of several excavations. The place was in the Sioux Reservation, fifteen miles from Antelope and was also known as the Black Horse Station.

Whitney Springs, seventeen miles further, was at the headwaters of Whitney Creek. Here in October of 1878 a tenderfoot driving a bull team had fourteen bulls stolen by four bandits. The other teamsters in the train became exceedingly riled and gave chase—and got the cattle back. The freighters were rough and tumble men and one thought a second time before tangling with them—as the bandits found out.

Twelve miles was Big Meadows where Northwestern built a station in April of 1878, at the same time the company built stations near the Moreau and between Crook City and the Belle Fourche, according to the Black Hills Daily Times.

At one time a terrible storm raged at Big Meadows and a man froze to death . . . not an uncommon occurrence when blizzards swept down from the icy north and played with the travelers as a cat does with a mouse. The smart pilgrims set up camp and waited out the storm; however, the freighters worried about their cattle and in checking them often fell victim to the storm's fury.

Robert Casey tells that one time Boone May and party came here on a buffalo hunt and had a battle with the Sioux. Six men were killed. One brave was wounded and May took a knife and "finished him off," without batting an eye. This area was prime buffalo country and Ed Lemmon told that in 1881 the Delaney Flats in Perkins County was swarming with the shaggy animals since it was on the Indian reservation and off limits to white hunters. In the fall Colonel James McLaughlin came here with Sitting Bull's Indians and killed five thousand buffalo—the last big buffalo killing in the United States. The slaughter ground covered about two townships and was mostly in Perkins County and a bit in Adams County in North Dakota.

The Bismarck Trail went east of the present town of Meadow—a small town with a post office and a handful of homes. The creek at the crossing was boggy and the early travelers filled the crossway with rocks—the crossing was about four to five miles west of the Foster Ranch.

The road continued to the southwest passing the present Chance on Thunder Butte Creek. Here are a couple of homes and a cemetery one-half mile away.

Rabbit Creek was crossed—a small stream that usually has water although it may run dry in the fall. At one time—in the homestead days—there was a town here called Ada. Today one house remains—the home of Mr. and Mrs. Odou Patterson. Ada grew when people arrived in large numbers to establish farms and ranches in the area, but with dry years, grasshoppers, low prices, hard winters—and even with a lot of hard work, not too many individuals "made it" and were forced to move away. Ada, as well as other settlements, "bit the dust," when the population dwindled. One of these settlements was the town of Moreau, established in 1900 by Ira Merrill, who died in Deadwood, a couple years ago. Ben Ash had a ranch at the conjunction of Rabbit and Moreau—Mr. Ash, you may remember, was one of the "fathers" of the Bismarck Road.

At Ada is a cement marker placed in honor of Spy Wilson, a freighter on the trail in 1876-77. Several of Wilson's daughters were present when the marker was erected, including Dorothy Dippert Lewton of Bison. The Butte County Historical Society has been instrumental in erecting signs marking the Bismarck Trail, as well as the stage stations from Crook City to the North Dakota line. The marker at Ada honors Spy Wilson, as well as marking the trail.

The route continued to the Moreau River where in early days there were many Indian raids. Nearby grew the town of Bixby, in homestead days, and when the modern highway was constructed the town moved to its present location. Towns in the old days, as well as today, are nourished by the current routes of travel.

Another fifteen miles was Cedar Canyon where the Black Hills could be sighted in the distance. This is picturesque country with deep canyons, cedar trees tucked in gullies and coulees—fine ranching country, mighty good antelope hunting country as well as a good recreational site. A man by the name of Donnenbeck ran a station approximately two miles west of the present Prairie Home Church and when he retired the station was moved about one mile to Cedar Canyon. There is a cement marker at the Prairie Home Bible Camp—about three-fourths of a mile from the Bismarck Trail. Not too far from here, on the Walter Gerbracht Ranch, are some rifle pits—whether a scene of a battle—or to prevent one—is a matter of conjecture. Mute testimony to the days when traveling was a risky venture.

Between Cedar Canyon and the next station—Frozenman—was the Gore Ranch, perhaps used in later days. It had a look-out tower, a large house, several small buildings and a large dugout barn. On the place was a stockade and a large pond—this according to Charles O'Donnell who stopped here sometime before 1911 and remembers that there were several men and two women at the stop—and the meals cost two bits!

The Frozenman Crossing, located on the present Calvin Stomprud Ranch, apparently had a way stop where the horses were changed and used by the stage in emergencies. Still to be seen are depressions of several buildings dug in the bank and above them is a trace of a house that reportedly was never completed.

The frozen man for whom the creek is named was an unfortunate Scotchman who was on the trail with a bull team and apparently alone which was rather unusual. A blizzard came up and claimed the life of the bullwhacker although his animals were later found on the prairie. The man's grave is about four and one-half miles from the crossing, it is marked with two sandstones, about the size of a football—one at the head and the other at the foot. A marker was up for many years that gave his initials and the date of his death.

This is only one of the numerous lone graves that are beside the roads into Deadwood—some victims of the weather as was this Scotchman, others killed by brigands or Indians, some whom met death in an accident or with some illness. To these unfortunate travelers, a grave on the prairie, usually now lost and unmarked, was the "end of the trail."

The next stop was at Sulphur on the present Mutchler Ranch—about twenty-seven miles from Newell on old 212. In the early 1900's the Jewett family lived here when the place was used as a stop by the homesteaders and traveling public. The house was situated on a hill and constructed of stone and logs. Across the road were large pole corrals. South and a little east of the house was a henhouse and a barn . . . by 1928 the old barn had fallen down. Jewett Creek runs on the south side of the buildings and toward the southeast. Southwest of the house was a well and across the highway was another shallow well for watering the stock.

Mrs. Walt Weaver of Whitewood, a stepdaughter of Mr. Jewett, remembers that the house faced south, there was one door to the dining room with the kitchen off this room. The other door from the entry way went to the living room and off this room were the bedrooms except for one on the other side of the house.

In the staging days the Sulphur Station was used for changing horses, in later years the mail stage from Newell came past. (The mail stage still comes past—only today it is a gas-powered vehicle.) Today there is nothing left at the station site except an occasional piece of glass or pottery.

Southeast of here was a town in homestead days called Acme, now long gone.

The next station was Alum Springs or Elm Springs that also was on old Highway 212 and not far from the Elvie Gotfredson Ranch. Today we wonder where there was water for travelers and learn that beneath the bridge is a water hole—and even in 1962 an extremely dry year, the ranchers were able to get water here. Early reference is made to springs and long ago trees grew along the stream bed . . . and on the Gotfredson Ranch is an artesian well with warm—and sulphurous flowing water. It is easy to see that people lived at the station site as remaining are pieces of broken glass, pottery, parts of a stove, pieces of dishes and square nails.

In 1922 a lady traveler stopped here for dinner and remembers one of the proprietor's children swinging a black and red sock that gained too much momentum and landed in someone's soup. She laughs remembering the incident and especially the look of dismay on the mother's face who, no doubt, felt like disowning that particular cherub.

In 1877 there were several attacks on the coaches in the vicinity, including one in November at the Sulphur Station, termed the first station beyond the Belle Fourche and the third beyond Crook. That would leave the Elm Station not yet established, but soon items mentioned the Sulphur Station as the second one beyond the Belle Fourche.

The pronunciation of Elm was often Alum by the early-day people and it was spelled that way too. It is difficult to know which station they were talking about—especially when there was an Alum Creek on the Miles City route. In fact there were Elm Creeks, Hay, Spring, Boxelder, Cottonwood and Willow Creeks on practically every trail to Deadwood. Thank goodness for such unique names as Cannonball and Louse Creeks!

From Alum or Elm Springs to the Belle Fourche was twelve miles of fairly flat country that went on the ridge above the Wallawa Ranch and through the Billy Uren Ranch. The Belle Fourche is a fair-sized stream, big enough to need a scow in the early days, one that was often out of order . . . and appropriately enough called the Mayflower. In windy weather the Mayflower wasn't to be trusted to get across and it was hoped to get a bridge in the summer of 1878. In 1879 a petition was circulated asking for a post office to be established here called Belle Creek. The residents complained that they had to go all the way to Crook City for their mail and this was sometimes very inconvenient as it was a good distance away.

The Belle Fourche Crossing made the news frequently—in November of 1877 a party was corralled here for four days due to Indian unrest. The following month a hunting party of six whites fought a band of Indians—the outcome of the fracas was not given. In 1880 a barn at the crossing burned and lost were one hundred tons of hay, all the grain and five horses. There was a rash of fires in the area, two or three each week and a firebug was suspected. The blame fell on a rancher, Ed Smith, because he had in his possession a trunk that had been at the station during one of the big fires . . . besides Ed was an unsavory character who had shot a man the previous season. He was cleared of the charge when it was decided that he had been too drunk to set any fires . . . but the matter of the trunk was not cleared up.

In 1880 General Sturgis informed the traders and others located at the Belle Fourche Crossing that they would have to move from one side of the river to the other . . . he claimed that the land on the east side was Indian reservation. However, maps showed he was mistaken as the Indian land did not approach the crossing by fifteen to eighteen miles.

When Bear Butte loomed in the distance the early travelers knew that their long trek was nearing the end. The mountain, a giant bear crouching at the outskirts of the Hills, took the early wayfarers three hours to get around it. Most of the travel was toward the south but many—especially freighters, took a shortcut to Crook City—the ruts still in evidence on unplowed land. One set of ruts can be seen on the Raymond Andersen ranch as well as neighboring ranches—this route may have been used by the mail carriers when mail was carried from Deadwood to Vale in 1884. The later Whitewood-Vale road closely followed this early trail, in fact in places the old trail can be seen on one side of the present road and the Vale-Whitewood road on the other. This trail went through the present Richmond Jones Ranch where it was joined by another shortcut road from the east. The trail went southwesterly and joined the stage road a short distance from the present town of Whitewood.

The Richmond Jones Ranch has a number of handsome buildings including a large house—the buildings almost in direct line with Bear Butte. East of the house was a livery barn and near the house was a well.

The main route went southerly via Spring Creek—one of the early ranches along the trail was Grasshopper Jim's place, the present home of Mr. and Mrs. Fred Secrest. Grasshopper Jim had come to the area in the very early days—he claimed he arrived in 1872. Indians were frequent early-day visitors nor were they always friendly. A stockade was constructed about two hundred yards from the present home, a fortification approximately fifty by one hundred feet and ten feet high. It was built with split logs with a trench about four feet deep—a sturdy stockade that undoubtedly would still be up if it hadn't been torn down some years ago.

At one time Jim was walled in by Indians in a nearby cave but his horses, also in the cave, didn't suffer one bit as hay and grain were stored there. Jim, however, got mighty hungry—the hay and grain weren't appealing, and to keep his strength up ate some grasshoppers and thus earned his nickname.

Indians were a constant menace in the area. In February of 1879 there were seventeen red warriors on the Fletcher Ranch eight miles from Crook City and half-way between the Bismarck and Sidney Roads. In January of 1878 a man was shot through the head on Jim's ranch, although this wasn't the work of the hostiles—it was thought to have been suicide. The man was William E. Wilson, known as Red Cloud, a freighter. The Wagnus family was killed by Indians near Bear Butte when they were on the way to Minnesota from Crook City. The men were not scalped but Mrs. Wagnus had her long, bright hair soon dangling on a brave's string of scalps.*

Near Jim's ranch was a tent town—the original Scooptown that grew up to house the hangers-on of the army encampment located here before Fort Meade was located. There was a commissary and liquor establishments and Billy Nuttal, from Deadwood, moved his theatre here. But drinks and theatre weren't enough for the soldiers who frequently came to Deadwood to do a bit of celebrating.

A few miles further was another stop for the freighters—at the James

Brown Ranch on Spring Creek and not far from Bear Butte. In August of 1879 he wrote about keeping a "cabbin hotel." This was in a letter to his wife who was back home in Iowa and he added, "It snowed yesterday and night before so the freighters have been flocking in for hay and meet and bread. Now dont that seem like business, since yesterday noon I have taken in $20 for those articles. Hay 75 cents per cwt. venison 5 to 7, head 25 cents per lodge dont you its business. You had aught what a town was here last night, four trains, 51 wagons, 104 horses and mules and 116 bulls and if you had been listening you might have thought there was a camp meeting over the way. There is only 3 small trains here tonight . . ."

Hay was twelve to fifteen dollars per load during the winter but most of the freighters quit for the season. In February a night herder who had been lost in a storm came in with sixty-nine head of cattle and Brown helped round up the stock. By the next season—1880—much of the freighting had gone to the Pierre route as Mr. Brown did not mention any more freighters stopping by.

An 1884 map shows a cutoff trail that left the main road a short distance from the Belle Fourche Crossing and went east of Bear Butte to Ford Meade. This is beautiful country east of the present Sturgis. The fort was constructed in 1878-79 and it is said General Sheridan who located the site went around it and pointed with his sabre where the various buildings were to be built. Later in 1878 it was mentioned that there was a hotel, other buildings, stores and a blacksmith shop under construction at Rhulon, the town near the fort. In 1880 the stage passengers would "break bulk" at Major Wilson's in Sturgis. At one time mention was made of a town called Saratoga that was located "at the mouth of Boulder Canyon."

In earlier days traffic from Sidney, Bismarck and Fort Pierre went via Crook City but in 1878 the Boulder Canyon road was opened and much of the traffic went through there.

Ruts of the old trails can be followed easily on virgin ground—but if the ground has been plowed the ruts have been erased. The Hills have ruts in only a few places . . . the topography here does not lend itself for retaining traces of the routes too well—however, in some places they are evident. In fact, one of the clearest ruts can be seen south of Nemo that were made over one hundred years ago by General George A. Custer and his men!

An interesting place to see the old Bismarck trail is on the August Nonnast farm south of Whitewood and north of Crook City . . . the tracks are especially noticeable in the fall and spring. But here, as on all the trails the traces are becoming ever more faint.

The first route from Bismarck was 211 miles—the road made a junction with the Sidney Stage Trail 2.34 miles from Crook City. From Crook City it was 4.85 miles to a junction with the Spearfish Road and Fort McKinney Road* and from there another 4.88 miles to Deadwood, these distances given by the Army Corps of Engineers. Later in 1877 the stage started going by the way of Fort Meade and the distance was then given as

242 miles.

*Some accounts say that the men—two Swedes were also scalped by Old Lame Deer. The bodies were brought to Crook City by Grasshopper Jim, a professional gambler, over six feet and always wore a white hat.

**There was a stop at Carstenson's Ranch site at Centennial City known also as the Valley Ranche. Some preferred staying here because it was quiet. At Crook City there was "too much shooting."

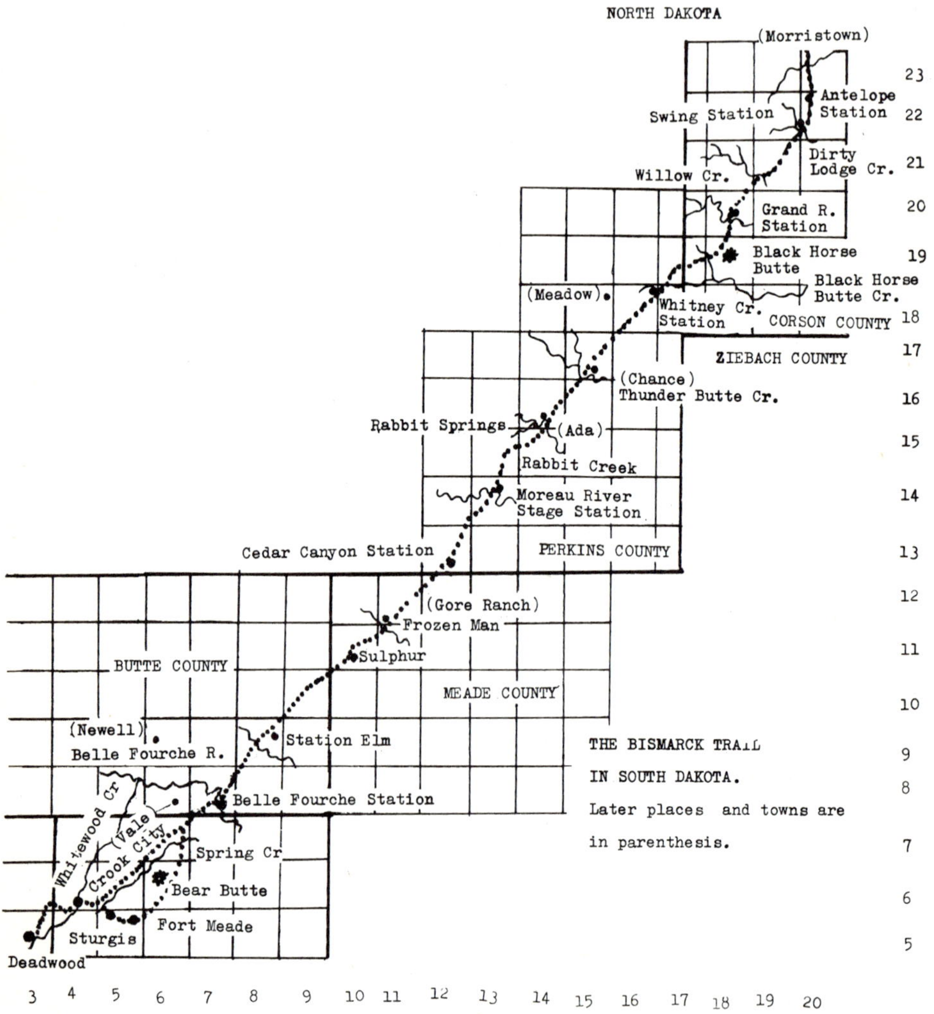


THE BISMARCK TRAIL IN SOUTH DAKOTA.

Later places and towns are in parenthesis.

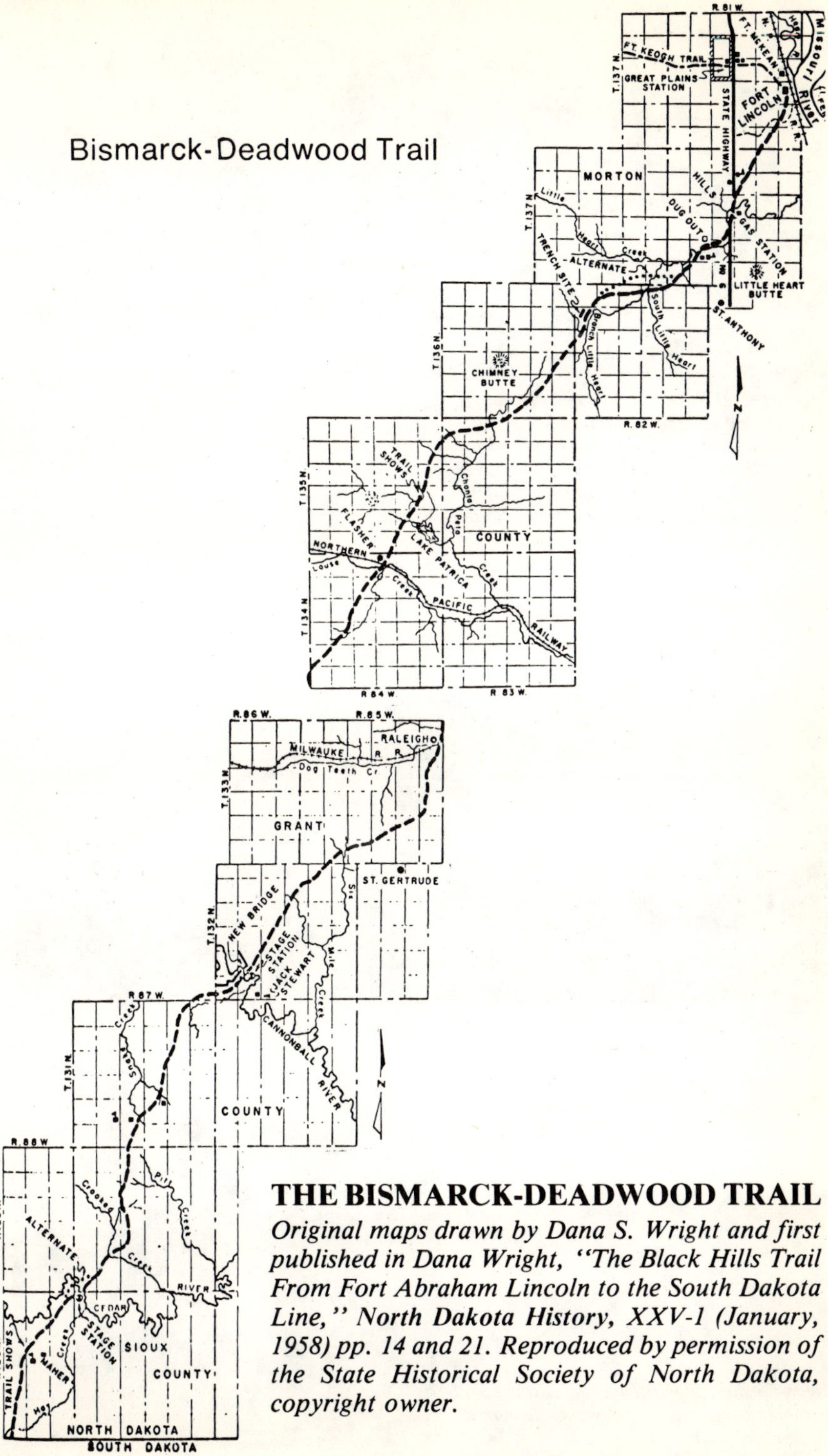


**THE BISMARCK-DEADWOOD TRAIL**

*Original maps drawn by Dana S. Wright and first published in Dana Wright, "The Black Hills Trail From Fort Abraham Lincoln to the South Dakota Line," North Dakota History, XXV-1 (January, 1958) pp. 14 and 21. Reproduced by permission of the State Historical Society of North Dakota, copyright owner.*

*Bismarck, D.T. in 1873 when it was a small town east of the Missouri River. Bismarck started growing in earnest when the Northern Pacific laid their rails here and the town became a jumping-off point to the Black Hills.—Courtesy State Historical Society of North Dakota.*

*How times have changed! In 1977 the state capitol is a skyscraper.*

*Near Flasher, North Dakota, is this sign commemorating the Deadwood-Bismarck Trail days.*

*There is a lovely picnic area at the Cannonball Crossing of the Bismarck Trail . . . a memorial to the old wayfarers and one that can be used by today's generation. The original buildings were a big horse barn, one small building for the station keeper and his wife, and a small building for supplies. Water was and still is from springs. Birds make their homes in the shelters and are alarmed at human intrusion.*

*Cedar Canyon in South Dakota . . . very picturesque country. The trail went along the edge of the area where traveling was easier. In 1915 the Cedar Canyon post office was still housed in a sod building.*

*Here at the Frozenman Crossing is the site of a house that was never finished. Shown are Mr. and Mrs. Calvin Stomprud and Kenny Wetz—on the Stomprud Ranch not far from here are deep ruts of the trail.*

*This shows the dugouts on the bank at the Frozenman Crossing. Many of the early buildings were dugouts or partial dugouts with the structures of sod or with sod roofs.*

*Marker in honor of Ben Ash and partners who blazed a trail from Bismarck to Custer in the Black Hills. The trail went a bit to the east of the marker.*

*The Prairie Home Church not far from the Ben Ash Monument . . . the stage line was a short distance from here.*

*The Sulphur Stage Station in the 1900's—Photo courtesy Mrs. Verla Jewett of Belle Fourche.*

*The Mutchlers standing at the site of the original Sulphur Stage Station—nothing is left here except a few bits of glass and pottery.*

*The site of the Belle Fourche Crossing. In the early days the crossing was often made on a scow called the Mayflower that was not a trustworthy conveyance—especially in windy weather.*

*Vernon Holst at the site of the Belle Fourche Crossing. Near here can be seen the excavations of the early dugouts. Vernon and the Butte County Historical Society have been instrumental in marking, with handsome cement markers the trail from Crook City to the North Dakota border. A few miles away is Vale that was established as a post office in July of 1883.*

*The Bismarck Trail went to the south much the same as this present road—the site near the Fred Secrest Ranch—the original home of Grasshopper Jim. Near here was an army encampment before Fort Meade was constructed.*

*This is the old trail north of Duane Carstenson's Ranch on Centennial Prairie. The road went toward the old Half-Way House and Deadwood. In 1893 a fierce forest fire raged here—one of the most devastating in the Black Hills.*

*William Selbie of Northwestern. He owned the land on which the present Whitewood is built.*

# 15
# THE MEDORA ROUTE

In April of 1884 the dashing Antoine de Vallambrosa, Marquis de Mores, came to Deadwood. The founder of the cattle town of Medora on the Little Missouri in the present North Dakota Badlands wanted to meet with the leading businessmen of the gulch city regarding a station on the Belle Fourche that would be on a freight and stage trail he was establishing from Medora to Deadwood. He planned to tap the rich freighting business in the Black Hills and the Deadwood tycoons gave him considerable encouragement. They already had lines from Pierre and from Sidney but they weren't too happy with them—and especially with the mail service that was far from being top-notch. They concluded that a line connecting with the Northern Pacific would be just great, and, besides a little competition might be a good thing.

Deadwood, in 1884, was past her rough frontier days, in fact she was putting on metropolitan airs. Stores had everything for sale that was sold in the States, there were several good hotels and restaurants, including the elegant IXL that served as good meals as did the Statler in Chicago! Electric lights were being installed and the Bank Saloon with the new-fangled lamps was noted as one of the handsomest resorts in the city. Telephones, however, were old hat—they had come in vogue in 1879. Yet—around the edges Deadwood retained a mining camp roughness; there were opium dens and saloons in abundance and plenty of gambling places. The narrow Main Street was muddy and there were big, medium and small potholes to vex the traveler.

Forty thousand people were living in the Black Hills and many mining companies with their mines and mills were producing gold and silver. There were the Caledonia, Fairview, DeSmet, Oro Fino, Iron Hill, Deadwood

Terra, Pecacho, Wooley, Chief of the Hills, the Esmerelda and the Aurora—to name just a few. The Sitting Bull in Galena, that had given a lot of business to the freighters, was closed, temporarily it was hoped. However, tin was becoming important in the Tinton Country and Carbonate, seven miles from Deadwood was a promising silver camp.

The Marquis, accompanied by Attorney Frank Allen and the freighters Pennell and Jones, also visited Lead where the Homestake Mining Company was expanding its operations and needed supplies and machinery for its mines and pounding mills. The superintendent of the company, H.M. Gregg, gave a tour of the giant works of the company to the visitors who were highly impressed. The four went back to Medora in good spirits and the Marquis gave the green light to Erasmus "Rassy" Deffenbach, an old Deadwoodite who had come to the camp in 1876, to get on with the route plans. With characteristic energy, Rassy, in company with Bob Roberts and Vic Smith, got busy and the Medora Route became a reality.

The people were fascinated with the French nobleman, a tall and slender swashbuckler, with dark eyes, a wax-tipped moustache; a man with a graceful carriage and who was tops with a sword or gun—a regular fire-eater ready to fight at the slap of a glove. He had charm and education and was a mite lordly with his neighboring ranchers which didn't set well with them. DeMores had come to America from France in 1883. He married the former Medora Von Hoffman and soon left for the west to do a bit of hunting. He was enchanted with the Badlands, a fantastic area along the Little Missouri—a land of weird domes, towers, buttes and ridges in shades of black, white, blue, reds and grays.

"Oui! Oui!" we can imagine the Marquis as saying, "This is the land! This will serve me well!"

The Frenchman had a vision of restoring the monarchy in his native land and this would require lots and lots of money. Here in this untamed region he had his answer—an ideal place for raising cattle. Feeding sheds could be set up and the steers fattened in two and one-half years instead of the usual five. He would slaughter them in his abbatoirs in Medora, Miles City and Billings and then ship directly to New York City in refrigerated cars. There he would set up fifty retail shops in the poorer section of the city where the meat would be sold from two to four cents less per pound than elsewhere. It was a good scheme but the man was ahead of his time, also, the rival packers weren't too enchanted with his plan. They undersold him at terrific losses and got rebates from the railroads and in time drove the Marquis out of business.

In the three years that the Marquis and his Medora lived in the Dakotas they left their mark. The Frenchman purchased 15,000 acres of rangeland, built a hotel, stores, a saloon and a picturesque Catholic Church. He constructed a club room complete with bowling facilities, erected a $300,000 slaughter house and built an exotic twenty-eight room chateau on a bluff overlooking the camp. Here the couple lived in grand style with a complement of servants, cooks, maids, valets and coachmen. They entertained visiting royalty and high society people who came out to see for themselves the "wild and wooly" west, the land of the cowboys. Occasionally a neighbor, Theodore "Old Four Eyes" Roosevelt of the nearby Elkhorn, Circle and Maltese spreads came visiting. Some claim that he and the Marquis didn't quite "hit it off," although Bill Smith, who rode for the Frenchman remembered seeing a cloud of dust coming along the prairie which turned out to be DeMores and Roosevelt in a buckboard—with a jug of whiskey companionably nestled between the two.

The Northern Pacific had laid rails west of Bismarck and was looking for a transshipping point to the Black Hills. DeMores thought it should be located in his town, as did Sandy Agno, one of the founders of Lead City, but there were other places in line for it—Gladstone, Belfield and Dickinson with its Short Pine Hills route. Dickinson, forty-five miles east of Medora appeared to be in the lead. In January of 1884 the citizens of that town raised money to place the road in top shape. Several bridges were built including a seventy-five foot span over the South Moreau that was fifty-five miles from Deadwood and one hundred miles from Dickinson—these figures appeared in the Black Hills Daily Times. Another bridge was constructed over Antelope Creek. Arrangements were made to have stages and express on this road and several Concord coaches were ordered but apparently didn't arrive because passengers went with the freighters. By March the Dickinson road was in full swing when teams with trails owned by Bickerdyke and Wamsley and of William Walton came in to Deadwood. A load of fruit was delivered to Central City merchants and a Sargent pulveriser for the Welcome Company in the Nevada Gulch area. Forty teams switched from the Pierre route when the freighters heard that 200,000 pounds of merchandise was waiting at Dickinson to be transported to the Hills. Seth Bullock and Alvin Fox went to the Belle Fourche to establish a city cutoff, called the Giles Cut-Off, that would save the freighters fifteen miles of travel.

The Marquis was having personal problems and couldn't do much to promote his Medora Route. He had been blackmailed and in standing off the men who confronted him, he shot a hunter by the name of Luffsey. The Frenchman was arrested, tried and acquitted; however, to get "even" his enemies poisoned six thousand head of sheep. (In the range country sheep were about as welcome as a maladjusted skunk.) As soon as the Marquis was freed, he turned his attention to the Medora Road that he maintained was a natural roadway between the Little Missouri and Missouri Rivers. He claimed it was only one hundred and seventy-five miles to Deadwood—a bit of underestimating. There was only one big stream to be bridged, said DeMores, this was the Belle Fourche. However, he didn't mention the thirty miles of gumbo in the Badlands that was the despair of the travelers in rainy weather.

The Belle Fourche didn't prove too much of an obstacle as the money for a bridge was donated by Minnesela, Butte County and Star and Bullock of Deadwood, who owned land on Redwater. This was evidently the bridge that was swept away by a break in the ice in 1886. Bullock selected the crossing, a site one-fourth of a mile above the mouth of Redwater where a seventy-five foot bridge was constructed that came to eight hundred dollars. Spearfish people raised six hundred dollars to bridge the smaller streams fifty miles beyond the Belle Fourche—and the Medora freight line started giving competition to the Dickinson Road. Several freighting outfits started using the Frenchman's route, including Ford with his seven mule outfit that brought in merchandise for the Fargo Store and S. P. Smith's bull train that came in with 24,000 pounds of freight; however, the trip took twenty-four days because of heavy roads. Bramble and Company thought the northern route might be a good one and they moved their books to Medora. It wasn't long before the shippers became most unhappy with this road—for one thing those thirty miles of murderous gumbo and secondly, there wasn't a warehouse in Medora. When the shippers were appraised of this they sent very little freight and it was noted that by July there were three hundred wagons in town—but no goods for them. The shippers had decided that Belfield, twenty miles to the east, would be a better and shorter route with superior accommodations—plus one of the best depots on the Northern Pacific line.

The Marquis offered to build a warehouse in Medora if business warranted it and the Northern Pacific was willing to contribute $2,500 toward the cost. The freighters said nothing doing—they wanted NO part of Medora and especially that thirty miles of glue and only an occasional freighter used the route. In July the trail went from bad to worse when bridges were washed away by summer "freshets." D. T. Bramble realized his mistake in promoting the Medora route and moved his office to Belfield. The Belfield Road to the Hills was improved. Twelve and one-half miles from the town the route intersected with the Medora Road and then followed the divide on the Little Missouri on the west and the Hart and Grand Rivers on the east. Thirty miles was S. T. Clark's Ranch; in 1884 he put up two hundred tons of hay to provide for the freighter's horses. He also stocked provisions and offered accommodations to the weary travelers. The South Moreau was crossed one hundred and twenty-five miles from Belfield and the next thirty-one miles was nice, rolling country up to the mouth of Whitewood Creek and from there was an easy road to Crook City and Deadwood, according to the Black Hills Daily Times.

This road was excellent in the wintertime as a rule because of less snow than on the other routes but it didn't help when the Deadwood freight office was closed during the cold months. One time in December of 1884 when a train arrived from Belfield only twelve days out, the air quivered with profanity when the teamsters discovered the office closed for the season. Customers also complained—they said it took twelve days for a letter to reach Deadwood from Belfield and it took ninety days for goods to come from New York City—and in that length of time even the styles would change. More and more freight went back to the Pierre route and then before too many years the railroad puffed into the Hills and traffic on the northern route was down to a trickle—mostly local hauling.

"Well, that's the way the cookie crumbles," said the Marquis regarding his freight route—or whatever expression was common at that time, and turned his energies to the stage road. As early as 1883 the Marquis had been thinking of a tri-weekly stage and with hopes of getting the mail contract. Several post offices came into existence between Medora and Minnesela, including Nashville, near the present Harding, that was mentioned as a new town in February of 1885, located in a pleasant valley in the Short Pine Hills ninety-five miles north of Deadwood. George Nash had the post office and Browning and Wringrose a store—both located in the same building. All these places were hoping for regular mail service when the Medora Stage Company got the mail contract—hopefully—but it ended that Northwestern underbid the Marquis and kept the mail on the Pierre Route.

Seth Bullock's town on the Belle Fourche was growing too, especially when the Marquis promised to make it a home station on his stage route. Seth reciprocated by naming the place DeMores. In May of 1884 surveyor White had completed his survey and plats and a hotel and four to five stores were assured of coming in. In June seven contracts for buildings were let and the stage company was putting up stables and an eating house. By July the town was fairly well established—enough to have the first ball in Jack Bishop's building. Forty-five couples were present at the affair, one more sedate than a cowboy dance held at Tugg's Saloon a bit later when every pane of glass was shot out. Mr. Tugg retired from the saloon business after only eleven days of dispensing "red-eye."

Bullock believed that the Northern Pacific would extend its lines to his town, as did the Marquis de Mores, and when it didn't Seth was instrumental in having the Elkhorn (Northwestern) come in from the south. When it did, it sounded the death knell for nearby Minnesela, especially when merchants there were offered free lots in the new town.

It was hoped to have the stages running by the 15th of September but the date had to be extended to October 1st. One hundred and fifty horses had been purchased at $90 to $125 each, some of them from Helena. When the animals arrived from the Montana city it was discovered that they were somewhat bruised from their long ride in railroad cars; however, they perked up in a few days. Four coaches were purchased from Gilmer and Salisbury, the former moguls of the Cheyenne-Black Hills Stage Company. The vehicles were repainted and U.S. Mail was stenciled in gilt and black lettering on the sides. They looked very handsome—almost as brand new. They had cost $1,200 secondhand; if new the cost would have been $1,500 each—an excellent bargain as Concords were sturdily built and lasted for years and years even after hard use on rutted prairie roads.

The Medora Stage and Forwarding Company still wasn't operational on the first as planned because there was work left to be done on the stations. Lumber, stoves and utensils were sent out. Tenders were placed at the stations—men who took care of the incoming teams and had the outgoing horses in harness and ready to hitch to the stage. They also took care of the harnesses and kept them gleaming. There were five real stations on the 192 mile route—real stations were considered the eating stops and the others were places for changing the horses. *The Black Hills Daily Times* and the *Badlands Cowboy* listed the stations as Medora, Davis Creek, Rocky Ridge, Springdale, Buffalo Creek, North Grand River, Cave Hills, Gallups, Middle Grand River, South Grand, Macys, Indian Creek, DeMores, Spearfish and Deadwood.

Finally the first Sunday of October the supply coach rolled into Deadwood and on October 7, 1884 the first coach—whether it was the Kittie, Medora, Dakota or Deadwood was not mentioned—left Deadwood and expected to reach Medora in thirty-six hours. The fare was $21.50, express ten cents a pound and the coach could carry 2500 pounds of express. Trips would be made tri-weekly. A driver would go a set distance to one of the stations, then double back with the coach coming from the opposite direction—his round trip would be about one hundred miles.

The stage company did not secure much passenger business nor forwarding business either although they had skilled reinsmen including Jim Welch and Bill Luce. Passenger comfort was looked after by providing buffalo robes in cold weather and the safety record of the company was tops. At no time did the coach tip over nor was there a single passenger accident . . . not even when during a period of time when there were six broken axles. It was discovered that some scoundrel was sawing the iron close to the shoulder and constant jostling loosened the axle. At one time when this happened the stage was on Centennial Prairie and the passengers were transferred to a sled and arrived safely in Deadwood.

The line kept losing money and in December was reorganized with a capital stock of $30,000 and with the Marquis DeMores as director. Some believed there would be additional traffic when the territorial capitol was moved to Bismarck from Yankton in January of 1885 but this was a false hope. The company reorganized again in April of 1885 with John Gaston in charge, but this didn't help either and in May the Medora-Deadwood stage was withdrawn. This didn't set well with the residents of DeMores and they considered changing the name of their town to one more "pleasant since the recent manifestations of the Marquis." The site of Demores, west of the town of today was eventually encircled by the town of Belle Fourche.

The Marquis was unhappy with the fate of the stage line but he wasn't one to sit back and moan about fate. He looked around for other ventures to keep Medora on the map and was highly interested when gold was discovered in the town—but only a small amount. More promising was the discovery of some clay and he considered putting up some pottery works, but this didn't materialize. The Frenchman still had his meat business, that is until someone instituted a whispering campaign hinting that there was something wrong with his frozen meat. This was the straw that broke the camel's back and the Marquis threw in the towel. He and his beautiful Medora went to India to do a bit of hunting. The Marquise was an excellent shot and before she left the Badlands her record for the year was two wild cats, grizzly and silver-tip bears, elk, deer, antelope and buffalo . . . and without a doubt she added to it in India.

The couple sailed for France and in his home country the Marquis experimented with a new kind of canned soup that he presented to the French government to supply the troops. He was involved in several duels . . . always the winner . . . he became interested in politics and became rather leftist. He went to Africa and was killed there under mysterious circumstances.

The Chateau in Medora was given to the North Dakota Historical Society and maintained as a tourist attraction with much of the money supplied by Harold Schafer of the Gold Seal Company, a multi-million dollar corporation. Schafer, a native of North Dakota, has renovated twenty-four buildings of the old ghost town, he and his wife, Sheila, produce a musical production based on the story of the Marquis and on Teddy Roosevelt, the program presented in a 1500 ampitheater every night from June 20th to August 31st.

So Medora lives again, a reminder of the days when cowboys clumped down the street, the cattle bawled when being herded into the abbatoir and the stage whirled in from Deadwood, the six-horse team stepping high. A memorial to those long ago days when the Marquis ruled his vast cattle empire, lived in regal splendor in his fashionable chateau and guided his stage and freight lines to the Hills of Gold.

## AHME CLEMENTS COMES TO DEADWOOD

An excellent picture of the Medora stage line can be gained from an account by Ahme Clements in a letter written to friends. The young lady was on her way to join her fiance, William L. Williams in Deadwood—this was in May of 1885. She boarded the train in the windy city of Chicago and started her westward trek and found the train "so smooth and runs so noiselessly that sometimes one can hardly tell whether the train is in motion or not." The meals aboard were very good.

Ahme mentioned the various stations, including LaCrosse, Red Wing, St. Paul and Minneapolis. St. Paul had a nice depot but it wasn't finished as yet—and the meals they served there were horrendous. Minneapolis she described as a burgeoning place with large flour mills, capable of making thirty thousand barrels of flour per day. Crossing the Minnesota border they passed little towns, including Jamestown, a very pretty place. Then the regular western country of prairies was entered—a boundless stretch of land that gave one "the same feeling as when viewing the ocean. There is a grandeur about it one would not believe without seeing it," declared Ahme.

Bismarck was a splendid looking town but as most western towns had an unfinished look. The Missouri here was one-half mile wide and called the Great Muddy. Out of Bismarck they hit a terrible blizzard—the worst one experienced in many years. At eight p.m. they arrived in Medora, a village with a few houses "dropped here and there as if by chance and with pigs running at large everywhere."

The young traveler checked at the stage station and was assured that bad weather or not, the coach would leave on time in the morning. She then walked over to the hotel which was a large unfinished building, exceedingly dirty and with a barroom on one side, chuck-full of elbow bending cowboys. She was shown to her room and found it had a stove . . . but no fire, there were a few chairs and a carpet that wasn't tacked down. Later she went down to the dining room and described it as an enormous place with room enough for one hundred people. The floor was black with mud and there was a red—and dirty—cloth on her table. Ahme lost her appetite and settled for a cup of tea.

Early in the morning the stage set out bravely with Ahme the only passenger if you didn't count her bird, no doubt, a canary that she had in a cage. The Badlands had terrible roads and it took three hours to go the first ten miles. The area was sometimes called Pyramid Park, a fantastic area with domes, mounds, pyramids and towers in various colors. There were meadowlarks singing sweetly, a few deer and antelope were to be seen and in the distance two shaggy buffalo were grazing.

At ten miles the horses were changed and Ahme then rode with the driver. At first she found this high perch made her dizzy but she soon got used to it and found it rode much easier. The driver was "Sitting Bull" Hall, an old Indian scout and very gentlemanly who never uttered a profane word in Ahme's hearing—not even when he had to get out every so often and dig the stage out of snow drifts in the coulees. Ahme became chilled and Sitting Bull gallantly wrapped his fur coat around her.

The second stop was an eating station run by a young man named Turner, who prepared a delicious meal with bacon, canned corn, prunes, bread and butter and coffee—but no milk. Mr. Turner kept the place, called Rockway, very neat—Ahme was exceedingly pleased after putting up with the dirty Medora Hotel. After eating the stage started out again but in six miles hit a blizzard and they had to turn back. Near the stop freighting men were camped, waiting out the storm. Each of them had to come to the station to look her over—members of the fair sex were in scant supply on the prairies. Ahme said she felt similar to "the remains at a well-conducted funeral."

Sitting Bull, Ahme and feathered friend started out at four in the morning and ran into a number of snowdrifts and overflowing streams. At one place Mr. Hall had to "pack over" his passenger, which meant that he carried Ahme on his back across the swollen stream. He then tried to drive the team across but the first two horses went down and the stage was stuck in the mud. The driver finally managed to get the horses across and they went on to the next stop where the horses were changed. This place was not a regular eating stop and the people there had nothing to offer for eats so Ahme shared her lunch—which was rather dry by then—with Sitting Bull. She waited here when the men went back to get the coach out of the mud and about five o'clock they were ready to set out again. A number of snowdrifts made for slow going and every few miles Sitting Bull would have to dig through a drift. At one especially big drift he took the horses and fastened them to the wheels of the stage as he said it would take several hours to clear the road. Ahme went to sleep but was awakened by a horse sticking his head through the curtains and she called Mr. Hall. He came back to the coach, fixed the curtains and assured his fair passenger that everything would be all right. Ahme couldn't sleep anymore, the horses acted so strange and wild. She would call out to Mr. Hall every little while and he would tell her not to worry. What a patient man! It was one-thirty before he opened the trail and Ahme begged him to come in the stage and rest, which he did by lying on the floor and in two minutes was fast asleep. At 3:30 the coach from Deadwood arrived, horses were changed, and Ahme, after saying her goodbyes to Mr. Hall, set off with the new driver, a man called Skinny. He said he hailed from Virginia and wasn't one bit as the first driver as he drank and swore a mile a minute.

At six o'clock they arrived at a home station that was fifty miles from Medora. Here a woman was in charge and served a nice breakfast of ham and eggs, hot cakes, cookies, bread and butter and with coffee to drink . . . cost of the meal was seventy-five cents. In jig time the stage was on the road again and at 9:30 p.m. they had made forty miles to a ranch called Case Hill* that had a post office.

The travelers started at eight in the morning and at 4:00 p.m. reached the next home station. A man was in charge here and served them a good dinner, including pie, although there was no milk. Ahme and the driver parted company but not before she gave him a stiff temperance lecture.

The horses were changed at intervals and Ahme warmed up at each stop. They came to a stream where the Deadwood coach was stuck in the mud . . . with the outfit was Mr. Allen, the general agent, who said he would see to it personally that Ahme got safely to her destination. The Medora-bound coach was freed and the south-bound coach with its six horses, driver and Ahme attempted to cross the stream but with no better luck. All the horses went down and water started coming in the door. Ahme was very frightened and Mr. Allen drove his buckboard alongside and our heroine took her bird cage in one hand, a satchel in the other and jumped aboard. She had to stay at the nearby stop, a very dirty place—in fact the dirtiest on the entire route. In the back of the building was a saloon with cowboys coming and going. One time when the door opened three little pigs came scampering in and had a great time running around the little sixteen by twenty-foot room. The bed, an Irish bloat type, was screened with a blanket fastened from the table leg to a chair. Ahme tried to sleep and managed to get a couple hours of shut-eye, if she had known a cowboy was posted as guard she might have rested easier. In the morning a man got breakfast but Ahme couldn't eat. Gradually the place filled up with cowboys and each ordered a dram of firewater for his breakfast.

When the stage was extricated from the stream it was damaged and

Ahme was heartsick thinking she would have to stay another day at this place. However, Mr. Allen volunteered to take her to Deadwood in his buckboard. Ahme gladly accepted the offer and was soon on the way to her William.

The newspaper in Deadwood gave her quite a write-up and complimented her saying she was a first-rate, plucky little lady. Ahme wrote to people back home that she liked Deadwood, that it "is so wild and such hills!" She and her new husband had a nice little house—with plastered walls even! And woman-like she said the stores in the mining camp were every bit as nice as those on National Avenue!

*Cave Hills?

# 16
# ON THE TRAIL FROM MEDORA
## From the Little Missouri to Deadwood

The *Badlands Cowboy* and the *Black Hills Daily Times* (who got the information from the Cowboy) listed the route of the Medora Stage as Davis Creek, Rocky Ridge, Springdale, Buffalo Creek, North Grand, Cave Hills, Gallups, Middle Grand, South Grand and Macys, Indian Creek, DeMores, Spearfish and Deadwood. Ten miles north from Indian Creek the Cowboy mentioned another station placed there in November of 1884—this was the Antelope Station.

It is odd that Gallup would be mentioned as it is shown on old maps as a creek that runs into the Little Missouri clear over in the western part of the present Harding County—the settlement of Gallup didn't come into existence until much later. There are several old-timers who believe the original trail may have been planned to go between the Cave Hills, across Gallup, Nashville near the present Harding and then to Macys and in 1885 the Black Hills Daily Times said that Nashville was the name of the new post office on the Medora Stage Road, ninety-five miles north of Deadwood. There are many others who maintain the route was the same as the freight trail—although there are forty-five miles between Macys and Bull Creek that we have no record of the early stops. This, of course, was in the early, early days—we know of the road ranches in later days—Korpelas for instance.

Crawford in his "Career of Marquis de Mores," Volume 13, Numbers 1 and 2, North Dakota History, gives the route as Davis Creek, Rocky Ridge, Robert's Spring, South Fork Cedar Creek, Cold Turkey Creek, North Grand, Crooked Creek, O'Dells, Bull Creek, Macy's, Antelope, Indian Creek, Belle Fourche, Spearfish and Deadwood. Crawford consulted old-timers who had lived during the Medora staging days and some who had worked for the Marquis so apparently he is correct—but that Gallup is still a puzzle.

The Medora Trail went south and east up Sully Creek and left it near the site of the George Pelissier Ranch. The first stop for changing horses was at the Davis Creek Crossing and the second stop was at Rocky Ridge where a Mr. Turner kept a neat place and offered good meals to the pilgrims—so said Ahme Clements in her story although she called the place Rockway. From there the trail went south to Robert's Spring (between Amidon and east of Chalky Buttes) although this stop was listed as Springdale by the Cowboy.

The trail continued to Cold Turkey Creek on the North Grand, according to Crawford, but the Cowboy listed the fifth stop as Buffalo Creek and the sixth as North Grand. Next was Crooked Creek, O'Dells and the Cowboy listed Cave Hills. (O'Dells could have been Cave Hills as Swan Gulch goes into the southeast corner of the Hills.)

Near O'Dell Alexander Connell later established a ranch . . . the trail went back of his place. In the early days he did freighting to the Black Hills and after his freighting days were over he settled on Bull Creek . . . north and a little east of the present Buffalo. Freighters used his place as a stop and he also provided mail service to his neighbors—the mail was thrown in a big box and people would rummage through it for any letters and papers addressed to them.

Connell was quite a character—Earl Clarkson of Belle Fourche remembers when his uncle introduced him to the tall bachelor, Alex remarked, "Clarkson?" then added thoughtfully, "I think there are more Clarksons here than white men." Earl had to agree that there were quite a few family members in the area at that time . . . this was in the early 1900's.

The trail went east of the present Buffalo, which is the county seat for Harding County and practically in the exact center of the county. About ten miles south of Buffalo was the Korpela Ranch that was a stop in homesteading days . . . there are no buildings left there now.

On the north fork of the Moreau was the Rice Horse Ranch . . . the owner sometimes had as many as a thousand horses for sale . . . many of them sold to the army. At that time there was a log house on the place and a barn with room for about ten horses. In 1900 the Van Horns purchased the ranch and the place was a stop for travelers who enjoyed Mrs. Van Horn's good cooking, nothing fancy but plenty of meat, potatoes and delicious homemade breads. Water was from a well and a spring.

Perhaps the most famous stop was Macy on the south fork of the Moreau. Ike Macy was an Indian scout, a buffalo hunter, freighter and a Civil War veteran who came to this area, forty-five miles from Belle Fourche—perhaps to rest from his previous exciting life. He established a ranch and built a log cabin that burned and was replaced in 1884 or 1885 with a two-story building that was remodeled in 1906. It was used as a post office for a good many years as well as a stop. The present barn was built about 1890.

Macy was on gumbo soil that produces abundantly with rain, which is sometimes in short supply in this part of the country. Three miles north is sandy soil—and fleas! The sandy soil makes for better traveling—and the gumbo, too, is fine—until it rains when it becomes something else. One day the stage had to wait out a rain but one woman passenger had no intention of just sitting around waiting for the trail to dry out and started afoot. She managed to go only a short distance when forced to turn back—her shoes rimmed with gumbo to the size of dishpans and her long black skirt heavy with mud. The woman had to agree with the stage driver that it would be best to wait for sunny skies and drying winds.

Ike was a character as was Connell—when the portly gentleman was asked if the snow in his house bothered him, answered, "Oh, no. It goes right through." When freighters were on their way for supplies he also had them get some for him and always gave them the final reminder, "And don't forget a quart of Yellowstone!"

Macy was a popular stop in the days of the homesteaders . . . prospects would come out from Belle Fourche and look over the land. They would stay overnight and in the morning look over the area, stay again that night and in the morning return to Belle Fourche. Some nights there would be thirty to forty people put up for the night—sleeping in the barn.

Ike's daughter, Mrs. B. Paul Baresch, now of Rapid City, remembers her Virginia-born mother baking up a storm "creating" pies and breads. A good customer for that bread was Indians who were inordinately fond of it—and any time a band passed by were sure to buy some.

Mrs. Macy religiously guarded one certain water hole that provided water for the household—in fact it had much better water than a well dug later.

Ten miles south of Macy was the Peck Ranch, used as a stop in later days. Another ten miles brought the travelers to the Peterson Ranch, the later Congdon place that was twenty miles north of the Belle Fourche. On the site, today, is a white ranch house, set among trees, a dam nearby and a number of neat outbuildings . . . the place owned by Melvin Delzer of Spearfish.

Three miles was the original stage station on Indian Creek, seventeen miles from the Belle Fourche, where Ed Williams ran the station. The water here is from surface wells—extremely hard water, yet a short distance away is soft water. The grass in the area is good for grazing. In later days the land was owned by the Hamiltons . . . it is now owned by Sarah and Harold Horman.

Across the road is a small hill with a marker at the top—the site of a lone grave. At one time there was another grave here too, of a twelve year old boy who died of pneumonia when his parents were camped here waiting for the creek waters to subside. The parents buried their son and went on their way. Later relatives disinterred the body and took it to their family cemetery. The remaining grave is of an infant, the child of a young couple who came here in the early '80's and the father got work with one of the local ranchers. Then one day some lawmen came to arrest him . . . there was some shooting but the man managed to escape and sometime during the night the woman joined him . . . but left a dead baby. If the infant was accidently smothered or otherwise, the rancher who had employed the father, never knew. He took a white ribbon, wrapped it around the tiny, lifeless form, placed the body in a box and buried it on a hill. Later he erected a small fence around the grave. In later years a gravel pit was dug here and the graves were moved closer to the Indian Creek Station. It was rumored that the fugitive had been a member of the Axelbee outlaw gang that had terrorized the Stoneville country in Montana.

There is nothing left at the station—only small bits of glass and pottery, rusted cans and several depressions in the ground showing the sites of buildings.

Ten miles south was the Owl Creek stop, no doubt only a watering place for the stage but used as a stop by the freighters and travelers. It is now the home of the Tom Davis family—in previous years there have been several owners, including the Brandts. Mrs. Brandt had only one hand but she could cook as well as any two-handed individual, even to cracking eggs and rolling out biscuits.

A dozen years ago the old barns were up but now all have been replaced as well as the house. There is a good well on the place—located in a different place of the early well.

The trail went on toward the Belle Fourche and about two miles from the river the roads separated—one fork crossed below the present diversion dam and the other a mile or so below at the Gilbert Crossing where there is a good rock bottom. These ruts lead toward Minnesela so apparently they were made in later days as the early trail went to DeMores. Out of DeMores the road went through a gap in Redwater Hill and it was here that Queen Ann was shot and killed. Queen Ann was the wife of a freighter, George King, and a rather high-falutin' and citified person—hence the royal sobriquet. When George died, Ann donned a buckskin suit and carried on with her husband's freighting business. One night she and two other freighters started out from Belle Fourche with Ann having as her cargo express packages and the payroll of a cattle company. On Redwater Hill she dropped behind her companions and when they came back to see what was holding her up, they found her shot, the wagon overturned, one horse dead and the other with a broken leg—and the payroll missing. Ann was buried in the cemetery at Crook City but for years people claimed that she came back to the scene of her demise—a white figure on the rimrock who wailed and screamed—lamenting whatever had occurred here years before.

The early road to Spearfish was further to the west than it is today and Spearfish was a small town when the Medora Stage came in. The stop was at the site of Kelley's Motel that burned a short time ago. The Medora Route came through Centennial Prairie, past the Half-Way House where Miss Harrington served "fault-less cuisine," over the Rest and down Deadwood Hill to the gulch city.

There were other stages in the north country after the days of the Medora stage—there was one from Harding and Camp Crook to Belle Fourche, and one from Bowman, North Dakota. Steven's had a stage line from Minnesela to Deadwood that was in operation for a number of years. A later-day stage driver was Red Percy who would give the tenderfeet quite a tale about the side-hill antelopes—jack rabbits. A freighter was Shorty Gardner who freighted with a six to eight horse team, driven with a rope and a gee and a haw. He had sideboards on his wagon, a tent for overnight sleeping, but would eat at Macys when in the vicinity—maybe a welcome change from his own cooking. Henry McCoid freighted supplies for his store at Harding—the store had been owned by O. O. Stokes, and before him the owner had been L. Harding. It was quite a store too—everything was for sale here—groceries, clothing, drugs, machinery—and coffins.

A lady who came to the area in 1885 found that mail service was non-existent as was passenger service. In the fall of that year Sarah Shevling left Kansas and when she arrived in Dickinson discovered that she had no way to get a letter to her sons in Burdick County—nor was there any transportation available for her to get there. She finally was able to hire a man to take her to Nashville, one hundred and thirty miles away. The two started out in an open buggy in the chilly November weather and there was not one house to be seen for forty miles until they arrived at the home of a Mrs. Malloch who lived in a small two-room building. Ten miles further Cap Olmstead lived in a comfortable house where the travelers stayed two days to wait out a storm. The clouds disappeared and they set out again and traveled forty miles to the Box Elder Cattle Camp; the only place they could stay from Olmsteads until they reached Ashcroft, eight miles from Camp Crook. Mrs. Shevling and driver had dinner at Ashcroft and then drove to Nashville where she was able to get in touch with her sons. (At that time there was no Harding County—the area was then part of Burdick County.)

Sarah kept a journal of events that happened in her life on the prairie and often complained of ill health. She mentioned a wagon load of provisions being taken from Camp Crook to Macy, the names of the early settlers, and her neighbors, including members of the Nash family, Mrs. Duncan, Dick Wells, Mr. Turbiville and Larimer Bill Freid. She told of the discovery of petroleum in the area, the prairie fires—and the rabbits eating up her garden. Mail came from Deadwood but in the fall of 1886 it was brought in by carrier with Ben Wannemaker the mailman. But most of all Mrs. Shevling wrote of the weather—balmy springs, hot summers and cold winters, the storms and the wind. The climate must have agreed with her—ill health or not—as she lived to a nice old age with a fat stack of journals to testify to it.

Perhaps traveling in the days of the Medora Stage when Ahme Clements was aboard or when Mrs. Shevling came from Dickinson by buggy sounds more romantic than traveling today, but I'm sure those sturdy individuals would opt for today's black-topped roads and comfortable vehicles instead of swaying coaches and open buggys, small buckboards or lumbering wagons that jolted over the gumbo trails on our northern plains.

***Medora sometime between 1883-1906 as the packing plant that burned in 1906 is in the photo.—Photo courtesy: State Historical Society of North Dakota***

*The Marquis De Mores who started the town of Medora. He started a meat packing industry and established a stage line to the Black Hills. The Marquis and Marquise lived in a handsome chalet at the outskirts of town, on a bluff overlooking the Little Missouri.—Photo courtesy State Historical Society of North Dakota.*

*The Medora Stage Line had four coaches —the Kittie, the Deadwood, Dakota and Medora—similar to the ones pictured in this famous photo by Grabill.—Photo courtesy: State Historical Society of North Dakota.*

*The stage barn at Macy, a famous stage and freight stop on the Medora and Dickinson routes. The barn was built about 1890—people often slept in it as in those days people usually carried their own bedding.—Photo courtesy: Mrs. B. Paul Baresch.*

*The house of the Macy family at Macy. The original log building burned and was replaced with this building in 1884-1885. —Photo courtesy Mrs. B. Paul Baresch*

*The site of the stage station on Indian Creek, on land now owned by Mr. and Mrs. Harold Horman. Mrs. Horman is on the right, the other lady is Mrs. Myra Hitzel of Belle Fourche.*

*The stage in front of the Billington Livery and Stage Barn in Spearfish—the stage came here from DeMores, the modern-day Belle Fourche.—Photo courtesy: Russell Whitlock.*

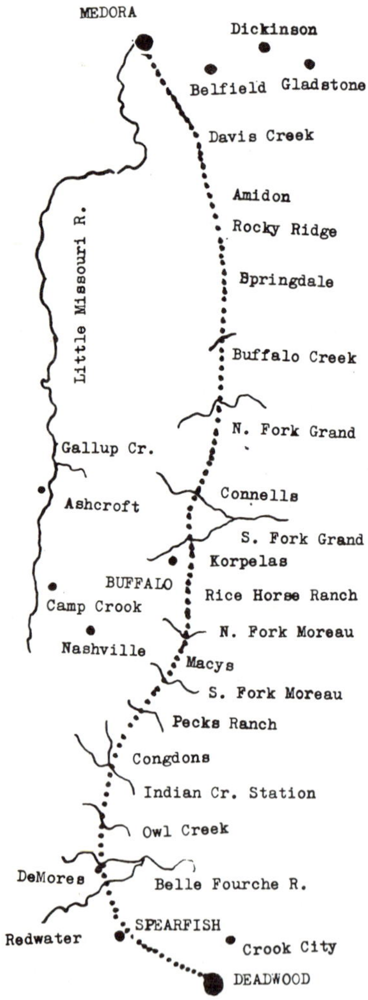


| CRAWFORD | THE BADLANDS COWBOY |
|---|---|
| Medora | Medora |
| Davis Creek | Davis Creek |
| Rocky Ridge | Rocky Ridge |
| Roberts Spring | Springdale |
| South Fork Cedar | Buffalo Creek |
| Cold Turkey Creek | North Grand |
| North Grand | Cave Hills |
| Crooked Creek | Gallups |
| O'Dells | Middle Grand |
| Bull Creek | South Grand |
| Macys | Macys |
| Antelope | Indian Creek |
| Indian Creek | DeMores |
| Belle Fourche | Spearfish |
| Spearfish | Deadwood |
| Deadwood | |

Jack Crawford gives the trail route in "Career of Marquis de Mores," Volume 13, Numbers 1 and 2, North Dakota History.

The Badlands Cowboy was the newspaper published in Medora. Stage stations were given in the October 2, 1884 edition. The same stations were listed in the Black Hills Daily Times published in Deadwood, South Dakota

In 1884, ten miles north of Indian Creek the Antelope Station was established.

THE MEDORA ROUTE

# 17
# THE MONTANA ROUTE

The Montana Trail can be mapped from Helena to Spearfish and Deadwood by historical places as guideposts—places with names that delight a poet, too. This was the first trail that "came from the west part-way on the Bozeman Trail, past Fort Phil Kearney, across the dry fork of the Powder River and leaving the plains by a canyon ten miles in length to the Belle Fourche. From there it crossed Inyan Kara Creek, over the Bear Lodge Mountains, past Sundance Mountain, twenty miles away from the 'burning coal fields' to Spearfish Valley and Deadwood."

This historic route was soon replaced—in June of 1876—by a short route that came via Miles City, Pumpkin Creek, Mizpah Creek, the Deadwood Crossing of the Powder River, Willow Creek, the Belle Fourche, Spearfish and Deadwood. Musical and poetic names, too.

Neither of these trails were used as widely as the other routes because the larger cities were further away from the Black Hills, the Indians were exceedingly active in Montana and there was a lack of railroads. The Union Pacific did reach Miles City in 1881 but the same line had already reached Bismarck, Dickinson, Belfield and Medora earlier and even they had bowed to Fort Pierre as the leading port, outfitter and jumping-off point to the gold country. Still the Montana Trail was used by individuals coming here from Helena, Miles City and other points and oxen trains came in as late as 1883. People also used the western routes when gold strikes were made at Judith Basin, Couer d'Alene and Alaska and the mail to Miles City and Alzada was sent from Deadwood in the early days and later from Spearfish, Whitewood, Minnesela and Belle Fourche.

Two of the first Montanans to arrive in the Hills were Charles Spencer and Edward Davis who arrived in 1875. In October they camped at the head of Whitewood and then went on to Custer. They may have left the Hills during the winter but they were back in the spring when they purchased Claim Number 6 below discovery on Whitewood—many of the Montana men were experienced miners.

The best known outfit that came from the west was the Montana Party that left Helena in March of 1876 with one hundred pack mules and a supply train and reached the vicinity of Spearfish in late May. These people were farmers, merchants and cowboys as well as miners—some of them stayed in Spearfish, including R. H. Evans who located a ranch one mile below the town, and his son, Alan and grandson, Billy, still live in Spearfish. Others went to the southern Hills including a group that worked Allen's Bar (seven miles below Hill City), where they took out six hundred dollars in gold dust, some of the men went to Rochford, Hill City and to other locations.

The majority of the members of the Montana Party and others who came in later appear to have settled near Deadwood where they did prospecting, some went in the freighting business, others started stores and various shops—including the Montana Bakery. William Bougey erected a livery and feed stable (forty by sixty feet!) in east Deadwood, John Haltiner became the milk man in Gayville and the Montana Corral and Livery Stable was operated by Job Lawrenson from 1876 to 1898 when it passed to M. B. Beckwith—who also had an ice business.

Burton and Cook took care of the horses of the Deadwood residents at a reasonable fee, pasturing them on Centennial Prairie near the old turn-off to Whitewood, making daily trips to the mining camp. During a raid the red men got away with approximately one hundred horses but fifty to sixty head got away and came thundering into Deadwood, wild-eyed and with manes flowing. The following spring Michael Burton went to Red Cloud to see if braves from Crazy Horse's band had surrendered any of the stock there—but no luck.

Two of the leading citizens in the Black Hills were Sol Starr and Seth Bullock who arrived from Helena in the summer of 1876. They shipped a load of hardware down the Missouri and then freighted it overland by wagon. The hardware was for a store in Deadwood and soon marked by one of the largest signs in the territory—easily seen miles and miles away—except for a few mountains in the way. They established branch stores in Sturgis, Sundance, Spearfish and Carbonate Camp, all handling stoves, wallpaper and furniture. One of their stove brands was the "Morning Glory" model that did everything but "start itself and cut its own firewood." A large choice of wallpapers was offered in the stacks and stacks of them piled practically to the ceiling. One time a customer had a narrow escape from an early death when a mountain of them fell on him. The new-fangled cupboards were something, too—they were supposed to keep the flies away from the food, one gentleman thought they weren't all that great—he maintained it would be too much bother to catch all those pesky buggers and put them in the cupboard!

Two other leading citizens from Montana were Ellis and McDonald who specialized in leather harness and saddlery. In November of 1879 they received a load of goods brought in by three teams with four trails drawn by twenty-four oxen—the biggest cattle ever seen in Deadwood.

One of the biggest names connected with the Miles City-Deadwood route was A. J. Maxwell, a rough, tough man who during his lifetime carved several notches in his gun although he maintained that a couple of the killings were accidental. One of the first notches was made in December of 1879 when Jack was taking supplies to his men at the mail stations between Miles City and Deadwood. At the Redwater Station he was joined by a man who didn't know Jack from Sitting Bull . . . and in those days one was mighty slow asking another's name. As they rode the man confided that he was William Bohrman, a deserter from the army. Jack asked him what he planned to do, Bohrman said he had several ideas in mind—and the first was to get some money. He had heard that a man named Maxwell was taking money to his mail riders and he planned to waylay him . . . and then "acquire" some stock at the stations and sell them at Bismarck and Sidney. "Say, Stranger, how about joining me in this little venture?" asked William. Jack hedged and said he would have to think it over . . . and admitted he had plans of his own.

When the two arrived at the mail shack on the Little Missouri Jack sent a telegram to Fort Keogh and said he had one of their deserters and to meet them at a cabin seven miles away where he would keep the man until the soldiers arrived. When Jack and William arrived at the designated cabin, Jack aimed his Sharp's rifle at the deserter and said, "I am Maxwell, will you take me in?"

"I'm a son of a b________ if I don't," replied the other and reached for his gun. Jack was about to tell his partner-of-the-trail to raise his hands when he touched the trigger of the gun, it fired and William Bohrman was instantly killed. Max gave himself up to Sheriff Bullock.

A few months later Jack was involved in another shooting and killed one of his drivers—again he claimed it was an accident. Evidentally he was exonerated both times but he had trouble again in 1883 when he had a fight with a black stock tender at the Box Elder Station. Maxwell evidentally got the worst of it as he had one ear and a finger bitten off. The tender was driven off and the paper nonchalantly surmised that the man, without a doubt, had been killed.

In 1880 Jack Maxwell placed an ad in the Yellowstone Journal stating that the Deadwood-Miles City Stage Line would commence running coaches on the first of April, connecting with all camps in the Hills. He had big plans to carry the mail daily and made arrangements for new coaches and accommodations for eight people—but the mail was still coming in three times a week in 1881 and the stage was a small buckboard in 1885.

In August of 1885 J. O. Hallowell was the new owner and manager of the Miles City to Deadwood stage line but there were no earthshaking changes except the mail was finally delivered tri-weekly—but only as far as

Spearfish. From there it was carried to Deadwood by Rogers and Spears. As late as 1894 there was express and passenger service to Miles City, provided by the Spearfish-Miles City Express which consisted of a small buckboard since business was far from booming.

Express service was also available to Sundance provided by the Spearfish and Sundance Express. Connections were made with Deadwood by the Spearfish and Deadwood Stage and Express Company—James Rogers, Prop.—this was in the summer of 1887. The outfit left Spearfish at 7:00 a.m. and 3 p.m. and left Deadwood at 8 a.m. and 2 p.m. A newspaper item in March of 1894 noted that Harvey Fellows, Spearfish hack driver, brought the mail in by pack horse due to heavy snow . . . one of the biggest snows in memory. Bald Mountain had ten feet of the "beautiful" and there was a generous covering of it all over the Hills. Harvey was the driver of the Spearfish-Deadwood Stage for many years; owner of the stage line was John McClintock.

Harvey often told of one of his passengers, a lady who came from Deadwood to visit a bachelor living near the present airport. The object of the visit was matrimony. The stage dropped her off . . . with her luggage. When Harvey returned from Spearfish, here was the same lady, with all her bags. One look at the bachelor and she realized he wasn't Prince Charming by the wildest stretch of imagination and she was happy to go back home.

The 235 miles of the Miles City-Deadwood and the later Miles City-Spearfish mail routes, in fact all of the mail roads in the area were some of the roughest stretches to be found anywhere. The carriers had terrible trails in the winter with frequent blizzards and below zero temperatures, in the spring they had to contend with roads that were mush "clear down to China." In the summer flash floods would turn quiet little streams—or dry creek beds—into raging monsters as in 1894 when Antone Gerig almost drowned in Willow Creek but did manage—finally—to get to high ground. The next day he recovered his buckboard but his team had drowned. Then to top it all, he was fined twenty-five dollars for bringing in the mail behind schedule!

One winter blizzard took the life of the seventeen year old mail carrier between Minnesela and Alzada—the youth was frozen to death. Joe Irvine, the Spearfish-Miles City mailman survived several storms by giving an oat-filled feedbag to his horses and he would take his buffalo coat and buffalo robe and burrow deep in a snow bank and wait out the storm cozy as the proverbial bug in a rug.

The Indians gave the traveling public a lot of trouble, frequently stealing horses as in September of 1880 all the horses were stolen at Pumpkin Creek. They also scalped a few individuals now and then. Nor were the road agents idle—in January of 1880 the stage was robbed at Box Elder about thirty-five miles west of the Little Missouri Telegraph Station—it was thought the bandit had been Badlands Charley.

In February of 1894 the driver of the Spearfish-Miles City Express Line was held up two miles below Spearfish. The night was almost as "bright as daylight'' and the driver was able to identify the holdup men as a black and a Mexican. The black—whose name was Frank White—was later arrested and lodged in the Deadwood jail—his feet and hands badly frosted. He was later freed because of insufficient evidence. The other, a man by the name of Lopez apparently got away free.

A bad man in the area—one with a penchant for stealing horses—was George Axelbee, who rode a magnificient gray stallion. His gang of outlaws headquartered at Stoneville (the later Alzada) ran things pretty much as they chose in southeastern Montana. In January of 1884 one of the gang members had been arrested in Miles City and was being brought to Dakota by Al Raymond, Deputy United States Marshall. To rescue the outlaw, his friends planned to pluck him from Raymond somewhere along the route. When the law heard of this Fred Williard and Jack O'Hara set out for Stoneville soon followed by A. M. Willard and McNarboe, a cowpuncher formerly with the Hash Knife outfit. The latter two arrived in Stoneville to find the outlaws at the saloon and Fred and Jack in Lew Stone's house. The gang members prepared to move out all the time calling the others vile names. The lawmen asked them to surrender. The gang refused and a heated-and-bloody battle broke out. Jack O'Hara, a deputy, was killed and Campbell and Axelbee wounded—Campbell was a member of the outlaws. Harry Tuthill, a friend of Axelbee, got on his mule and was also shot.

A group of cowboys were in front of the saloon watching the fireworks and the deputies thought these men were shooting at them, too—it turned out that an outlaw by the name of Billy McCarthy (Billy the Kid) had mingled with the cowmen and was taking potshots toward Stone's house. Two bystanders, cowboys, were shot by Axelbee who had a grudge against them.

Axelbee got away but this broke up the gang and he and one of his men were killed about four months later. Harry Tuthill was taken to the Spearfish Hospital to be treated, after about two weeks six men came in one night, dragged him to the edge of town and hanged him. It was learned that Harry was of a prominent New York family and very spirited in his youth. He was sent to Iowa where he grew up and as a young man came west. At first he hunted buffalo and then turned to horse trading and evidently did quite a bit of business with the Axelbee Gang although he apparently didn't engage in any horse stealing himself.

A famous individual who rode many trails to Deadwood—and she likely rode the Miles City route, too—was Calamity Jane. In 1873 Calamity and her infant daughter left Livingston, headed for the east, but not by Deadwood as it wasn't on the map at that time. She rode ahead of the wagon on her horse and did handstands in the saddle, the Indians thought she was "sick in the head," and permitted the party to go through unharmed.

Most of the Hills people have heard rumors that she was married to Wild Bill Hickok and that the couple had a daughter. It may or may not have been true and some claim that this daughter was the infant Calamity brought through hostile Indian country. At any rate the Department of

Welfare believed the evidence and deemed Mrs. Jean Hickok McCormick eligible for assistance with her birthday fixed as September 25, 1873 as given in Calamity's dairy and other papers. But Hills people still insist that Calamity may have been married to a Hickok but it was a relative of Bill—not Bill himself.

There is still traffic on the Miles City-Black Hills roads, now the vehicles are gas-powered. There is no longer need for the rules given to passengers on stages as given below—and on what route it was posted we don't know. But it is good for a chuckle or two . . . even if you have seen it before.

1. Abstinence from liquor is requested, but if you must drink, share the bottle. To do otherwise makes you appear selfish and un-neighborly.

2. If ladies are present, gentlemen are urged to forego smoking cigars and pipes as the odor of same is repugnant to the Gentle Sex. Chewing tobacco is permitted, but spit WITH the wind, and not against it.

3. Gentlemen must refrain from the use of rough language in the presence of Ladies and Children.

4. Buffalo robes are provided for your comfort during cold weather. Hogging robes will not be tolerated and the offender will be made to ride with the driver.

5. Don't snore loudly while sleeping or use your fellow passenger's shoulder for a pillow; he (or she) may not understand and friction may result.

6. Firearms may be kept on your person for use in emergencies. Do not fire them for pleasure or shoot at wild animals as the sound riles the horses.

7. In the event of runaway horses, remain calm. Leaping from the coach in panic will leave you injured, at the mercy of the elements, hostile Indians or hungry coyotes.

8. Forbidden topics of discussion are Stagecoach robberies and Indian uprisings.

9. Gents guilty of unchivalrous behavior toward Lady Passengers will be put off the stage. It's a long walk back. A word to the Wise is sufficient.

# 18
# ON THE TRAIL FROM MILES CITY
## From the Yellowstone to Deadwood City

Fort Keogh, that mothered Miles City, was located in 1876 where the Tongue and Yellowstone meet in Montana and at first was a collection of dark and dismal huts infested with vermin. But that winter, the troops under Colonel Nelson A. Miles, saw very little of their quarters as their commander kept them on the field a great deal of the time "chastizing" the Indians. The post was soon built up and by the end of 1877 had a number of handsome buildings including a library and a billiard hall, the buildings placed in the shape of a diamond. The fort, that had no stockade, was well constructed but not as beautiful as others in the state.

The settlement of Milestown sprang up next to one of the fort's barracks and stayed in that location when the post was moved one mile away. Milestown became Miles City and in time grew to be an important cattle and horse center, located on prairie lands that had fantastic herds of buffalo in the early days . . . in 1880 it was estimated that there were 30,000 of the shaggy beasts in the area. The people discovered that this was super country for raising beef cattle and the buffalo were dispatched to eternity to make room for the cattle. One time some men spotted objects that looked as moving haystacks and closer inspection revealed them to be wagons loaded with tens and tens of buffalo hides on the way to the Yellowstone to catch boats going down-river. It was then that Bill Cody received his nickname when he was a guide for the Grand Duke Alexis who came here on a buffalo hunt and Cody killed so many of the animals that the duke gave him the "monicker" of Buffalo Bill that stuck with him the rest of his life. With the buffalo out of the way—many left on the prairie to rot—great herds of cattle were trailed in from Texas and some from the West coast as well.

The fort, cattle, horses, the Northern Pacific—all helped Miles City to grow. In March of 1878 it was noted that the place had a store that "never takes in less than one hundred dollars a day," and there were five other stores doing well, there were ten saloons doing even better and, of course, there were several gambling places doing a flourishing business. In one of the saloons, Madam Bulldog, lately of Deadwood, was dealing "stud horse poker." The residents of frontier towns liked games of chance and Miles City was no exception.

In 1879 the newspaper reported that during the months of June, July, August and September, Miles City had grown considerably—on several streets whole rows of houses were standing where in May there had been only empty lots. Frame buildings had taken over the original log buildings and the cross streets were filling up fast with neat edifices. In fact the Yellowstone Journal was happy to report that their stores would be "creditable in towns of several times the size."

Although Miles City was a burgeoning place there wasn't much traffic to Deadwood; however, there was mail service for many years, first to Deadwood, then Spearfish and finally Minnesela and Belle Fourche. As mentioned the "stage" was a buckboard for most of the time . . . at least to Spearfish and Deadwood.

In 1879 a telegraph line was built from Fort Keogh to Fort Meade and to Fort Sully on the Missouri. The poles were of green wood and in a short time deteriorated and by 1885 the line was no longer in use. Before then communication between army posts had been by heliograph (or mirrors) from one high butte to another—and there are plenty of picturesque buttes in this country! The early stage line to the Black Hills closely followed the early telegraph line.

The trail left Miles City by following the line for a distance, modern 312 also follows it fairly closely. Seventeen miles from Miles City was a stop at the Gray Dutch Inn, the site of the present Houston Rogers Ranch.

From the Gray Dutch Inn the road generally followed the present one except the old trail made two crossings of the Pumpkin in Sections 17 and 21. As the road turns east on the Powderville Road we are at the site of the Beebee post office. The old log livery barn is still standing but the town is gone that was on the opposite side of the road although remaining are traces of buildings, excavations, parts of stoves, and bits of crockery and glass.

East of the Beebee Station is Deer Creek but the old stage trail did not follow the Deer Creek Valley but went due south for a mile and then eastward . . . the present road does use the valley. The stage station was about six to seven miles from Beebee, located near an excellent spring, on a ridge or 'hogsback" of the divide between Pine and High Creeks. Mrs. Payne, an early-comer to the area, remembered in later years, that on her first trip she had a delicious dinner at this station that was noted as one of the best on the line. Afterwards she went on to Powderville on the stage and thought everything was so romantic on that moonlight ride—perhaps as a young bride she saw everything in a happy, pink glow. A later stop was located on the present TONN Ranch with very little to mark it now—even the pole corrals are gone that were up for many years.

A short distance from the station is the present L.O. Ranch that appears as a small town with many buildings including houses, barns, sheds and an old schoolhouse . . . the ranch was a good stopping place for travelers in later days. J. M. Holt came to the Montana country in the early days from northwest Missouri and built up a huge spread here with his cattle ranging from the Chalk Buttes to the Powder and Little Powder Rivers and the Tongue River. Mrs. Holt lived in Miles City but there were men cooks on the ranch and they could whip up a good meal in short order. On the place were bunk houses, a commissary—and plenty of good western hospitality. The L.O. was sold in the thirties and it is said that the son who sold it slipped in a section that no one knew about.

Some sources give a stage stop at the confluence of Sand Creek and the Mizpah and an old newspaper article mentions that there was a bad fire here in 1887 that burned the station as well as several homes, barns, the store and post office. Reverend Eugene Szalay, who did a lot of work in tracing the old trail, mentions this area as the major part of the route and gave the route as coming to the Sand Creek Crossing where there was a corral and Holt's Gardens and the start of the telegraph line going north. After crossing Sand Creek, the trail went south and passed Maxwell's house just before crossing the Mizpah . . . this was according to an 1885 survey of the area.

The old trail follows the present road fairly closely all the way to Powderville. There was a relay station at the present site near the Ed Orestad Ranch on Spring Creek and nine miles from Powderville. There are some excavations in a bank and a foundation to mark the place and many years ago Mr. Orestad found a whetstone left by an early resident. Spring Creek runs all year and was a favorite place for bears in the early days.

The stage trail crossed the Powder River in the southeast quarter of section thirteen, at the site of the old Powderville post office—which has been moved many times. The place was known as the Deadwood Crossing at one time and then as Elkhorn—in the '70's there was a telegraph station here. There are several explanations how the town got the name of Powderville—one of them was given by Fred Henning who came to the area in 1879. He maintained that one day when there were several cowboys having a sociable time in the log saloon someone shot into a keg of powder and the cowboy sitting on the keg and the roof of the building went sky-high. Hank, the cowboy who went flying so unexpectedly, became a cripple for the remainder of his life . . . and Powderville got its name. (We thank Waldo Lewis of Miles City for this story . . . he was a friend of Mr. Henning many years ago.)

The crossing became an important mail distribution center for the people living up and down the river and its tributaries. When the river was high a flat boat that had been brought in from the Missouri River was used to cross the waters. Orestad, who grew up on a ranch three miles north of Powderville, remembers a man named Heine Gross living in a dugout at the crossing. Later he learned that the dugout had been used as a stop for the

Deadwood-Miles City Stage people at one time. Septon "Zipp" Cady was the assistant postmaster here . . . about 1883. He told that his main job was having charge of the horses and to have them ready when the stage pulled in, as very little time was spent at the stations. Then, as now, the "mail must go through" and Zipp did his part.

In 1883 Mrs. Robert Bowman who had had charge of the station returned to Deadwood.

As in many of the frontier towns, there was a lot of shooting as well as gambling and drinking with many of the participants of these activities now lying in the local Boothill above the camp—including one Deadwood Dick. There were many Deadwood Dicks in the west and this was Montana's contribution to the list. Not that everyone lived a wild, shoot- 'em-up life—the William Ferdons, originally of Washington, D.C. attended their evening meals in full dress—ties and tails for the men and decollette gowns for the ladies.

The post office is now located in a small white building next to the highway and is open three days a week. Near the old crossing is the Dean Keil Ranch on the west side and the Tom Creston Ranch on the east side. From the Creston place the road turned southward, as does the present road. It picked up the Crow Creek watershed to the west, there are some modern variations to the road; however, the line of travel is southward. Where the Pocochedie joins the Crow from the east was a stage station, four and one-half miles from the present Pete Hill Ranch where Hollyhock Annie ran the stop. Annie offered bacon (often on the rancid side) as the piece de resistance to the wayfarers—a staple meat on the frontier. She was especially fond of hollyhocks—we can picture the sturdy flowers growing at the doorway, perhaps watered with the dish water after the dishes had been washed and dried. It could be that Annie had a few cats around to keep her company—perhaps a dog or two as this is wide country with few trees to echo the loneliness. But there are plenty of hills with rounded tops cut by coulees and dry creeks. There is very little to show where Hollyhock Annie had her place—a couple depressions in the ground and that is all.

In later years many travelers stopped at the Jensen Ranch, about one mile from the present Pete Hill home. Mrs. Hill's father came to the country in 1894 and built a three bedroom house with a sleeping porch and bunk houses so there was plenty of sleeping places for the travelers. On the ranch was a store house, the grocery store was miles and miles away, hence supplies—especially for the winter months were a necessity.

Following the trail today it is hard to trace it in the area where Dry Creek joins the Crow but it did cross the Crow as does the modern road and stayed on the west side. This is an area of buttes and ridges with rough terrain but the road kept going on a southerly course. On the divide between Crook Creek and the Dry Creek Breaks was a stop northwest of the Bill William's Ranch. The site went through a gap in a range of hills into a valley where there was a good spring. In the late 1920's and '30's Mrs. Vera Lambert, now of Belle Fourche, came here picking berries and remembers that there were only depressions where the barn had been on one side of the draw and the house on the opposite side. The site was north of the present Hammond.

At one time Roger Williams found a gun in the area which may have been used in the robbery of the stage "thirty miles west of the Little Missouri Station"—this according to an early newspaper account. The holdup artist was a tough hombre by the name of Charles Davis, alias Badlands Charley who was eventually hanged.

The road went to the Box Elder where there was a station on the banks of this creek that usually has water but in the summer does go down to a mere trickle. On the banks grow ash, boxelder and a few cottonwoods—in the lowlands and surrounding area timothy grasses wave in summer breezes. Timothy was not native to the country, it was brought in by early freighters who often brought in their own hay and the seeds established it here.

The first building at the stop was constructed of logs with two bedrooms, a kitchen and pantry. Nearby was a log barn but one day a cyclone hit the place and demolished the barn and took the roof off one of the bedrooms. In time a second house was built, replaced by a third one—the home today of Mr. and Mrs. Lester Phillippi.

The Box Elder Station was north and a bit east of Hammond.

Willow Creek was the next creek crossed by the Miles City Trail, it then went to the present-day Alzada on the Little Missouri. The place was first called Telegraph Point with the telegraph "office" in a dugout . . . at one time a man named Schruster was the operator here . . . his main job to report the movements of the Indians (there were plenty of them here) to Fort Meade or Fort Keogh personnel. Poor Schruster was terrified of the red men, no doubt his teeth chattered as fast as the telegraph key many a time.

Lew Stone laid out a town near the site—in 1882 it was noted that the place was "complete with avenues, boulevards, parks and a fountain!" Stone ran the combined store and post office. In 1882 the "little Pilgrim camp" on the Little Missouri had ten families and was headquarters for hunters and trappers. In 1890 the town's name was changed to Alzada in honor of Alzada Sheldon, wife of Ezekiel B. Sheldon, who had come to the area in 1883 . . . seems that "Stoneville" conflicted with a town of the same name in Dakota.

The original town was a bit west of the present one and the telegraph dugout even further to the west. The first village was located near a swale that made it hard for the travelers—especially in wet weather. At the location can be seen the site of a dugout barn and other buildings and there is one building almost tumbled down.

The Little Missouri Crossing and Stoneville frequently made the papers—for instance in August of 1880 prairie fires were reported in the vicinity. At one time there were thirty-three Indians at the crossing and Frank Gray, the mail carrier, was concerned when one of them followed him for six miles. An Indian was thought to have killed two trappers and stolen their horses, no wonder Gray was more than a mite uneasy. At one time George Schoener's horses were auctioned off at Alzada, bringing fifty cents to $2.50—a pitiful amount. George was the Miles City stage driver and we can't help but wonder what he then did for transportation.

From Stoneville the trail went toward the Belle Fourche River that makes a triangle in its course through the area, with the northern-most point south and a bit west of Alzada. There were several crossings with one of the best being the Baxter Crossing at Riversdale, northeast of the present Keith McDonald Ranch and seven miles from Colony in Wyoming. An 1889 map shows the Miles City road following the river on the east side with one crossing above Riversdale and another at the settlement.

Riversdale was located where Pine Creek joins the Belle Fourche—here Bate Baxter had the road ranch, with a post office, in a long, log building. There was also a store in the hamlet. Carl LeSuer had a good artesian well on his land—there is nothing to show where the small town was located—no store, no ranch—not even the well. Today the site is owned by Thorval Jensen—next to it is land owned by the Mule Shoe.

Francis McDonald, who lived in the vicinity as a boy, tells of another trail that crossed the Belle Fourche several miles west of the Baxter Crossing, went past their ranch, past the present Keith McDonald Ranch and then toward a gap in a line of hills to the southeast where Crow Peak stands sentinel in the far distance. It is possible that this was one of the Miles City roads—it also could have been used in later days by people going to Mona, a town that has now "bit the dust," as have many of these early settlements. In those days there were no "laid-out" roads and no fences—hence there were trails everywhere on the prairies, many of which can still be followed by the ruts engraved on the land. In fact, some of these early routes became the modern roads of today, some were used by the homesteaders and others by the mail carriers—at one time there was a post office in the McDonald home. (We still have mail carriers except the mail stages are gas-powered and travel the routes in a matter of hours where it used to take days.)

The Miles City Trail crossed Pine Creek, Alum Creek and approaching the present John Moline Ranch was a relay station—no evidence of it remains. John has erected a rough sign pointing to the station site with an added message, "No Services!" At the Moline Ranch on Oak Creek was a watering stop with the well built into the bank of the stream. It had excellent water. From here the trail crossed the present road, went to the southeast toward the early Bunney Ranch.

Reverend Szalay gave the route as Red Oak Spring, Red Oak Creek, the Sourdough Road, Hay Creek, the Breaks of Redwater, McNenny, Spearfish Valley and Spearfish. Incidentally that interesting name of Sourdough Road got its name from a sourdough biscuit—delectable items made by a certain lady in the vicinity. So well known were her biscuits that not only was the road named in their honor but the flat where she lived became known as Sourdough Flat. The road takes the area people to Belle Fourche.

The Miles City road went over a gap into the valley of Hay Creek—and where the several creeks joined was the Forks where today is located the Chris Helmer Ranch. Here is an old building used as a way stop in later days—the place set within towering trees. At the Forks was a post office located in various buildings at different times—two of the postmasters were Ward Brown and Mrs. Andrew (Emma) Olson. The Helmer Ranch is near the old railroad that went from Belle Fourche to the Hay Creek "coal banks"—the present Aladdin. In 1880 the Miles City Mail Stage changed the route to go by the way of the coal camps.

From the Helmer Ranch the trail went to the east and a few miles from their place was the James Rathburn Ranch that was a stage stop. James Rathburn settled here in the late '70's where he constructed a snug, log house and frame barn . . . later he enlarged the house and covered it with siding. Both buildings are now gone.

When Antone Gerig was the mail carrier he left a change of horses here at the ranch . . . and, no doubt, frequently sampled Mrs. Rathburn's good cooking. She often served ham—cured in their own smokehouse and sauerkraut from cabbage raised in their garden. It is said that in later days the Rathburns did not charge for meals when anyone stopped by—good western hospitality at its best.

James had a peg-leg, the result of a buffalo hunting trip with some friends when he froze his foot—and twenty years later finally got around to having it amputated and fitted with a wooden peg, which in time was replaced with a modern prosthesis. But peg-leg or not, Mr. Rathburn rode horses, built up a thriving ranch and even hauled coal from the Aladdin area.

These early settlers were sturdy individuals, including Frank Cody (Keith McDonald's grandfather), James Rathburn and a man named Breckenridge who may have been the hardiest one of all. He came to northwestern Wyoming even earlier than Rathburn—long before 1876. The government took a dim view of these early settlers and tried to keep them out . . . or if they managed to elude the soldiers were taken out when found. At one time some soldiers escorted Breckenridge to Fort Pierre and had him walk the entire distance with his hands tied behind his back. This didn't faze the frontier man one bit—he was back on Oak Creek in no time flat. Eventually the land was opened for settlement, but Breckenridge couldn't stand having neighbors so close and went to the Grand River and the Moreau and then to Canada. Shades of Daniel Boone who couldn't stand neighbors within a hundred miles or so!

From the Rathburns the old trail veered toward the south over gently rolling hills, through the Glen Riley Ranch, Robert McKay land, the old Bronson place, the old Cunningham Ranch and past the Redwater School. At the top of the gap there is a magnificent breathtaking view of Redwater Valley with Crow Peak to the south. Red rocks jut out of the ground—carved by winds of centuries into various shapes and forms. The Redwater School is gone—just the foundation remains. Redwater is the line between Butte County and Lawrence County—the trail crossed on the present

Thompson Ranch where there was a good rock bottom. Today there is more water in the river than in the old days because there are three to four artesian wells that go into Crow Creek that in turn empties into Redwater.

The road crossed Lake Creek on Ralph Bahn's land—here is an excellent crossing with another a short distance away. Next to the Bahn Crossing is a venerable ash tree said to have been a "hanging tree" for some unfortunate individual who ended his life dangling from one of the branches—perhaps a horse thief—but that is only a guess. However, horse thieves decorated trees more than other types of low-life—and Spearfish had several hangings in the area.

Lake Creek runs from Cox's Lake—the little body of water that stage passengers were told had no bottom; however, in reality it is a spring with the bottom about sixty feet below.

The Bahn place was settled in the early days—thought to have been in the late '70's, by a man named McAleer, who filed the first water right in the Deadwood courthouse. On the Bahn place—likely built by McAleer—is an old, picturesque cabin standing on its "last logs," and nearby is an equally interesting spring house with a moss covered roof.

From the Bahns the road went on the old Morris Walsh place that is now owned by the Laussers—here is a stone barn built at the turn of the century. Then the trail went on the Melvin and Carl Anderson Ranch where the route can easily be followed through their pasture land. Next to the Andersons is the old Wahl place—Mr. Wahl was an early day rancher in the area who often talked about the old Miles City Trail when it was busy with traffic—freighters, mail carriers, travelers of all kinds, including cowboys who were loud and noisy going on an "R and R" trip to Spearfish or Deadwood—and sometimes even noisier on the way back.

On Wahl land by the side of the old trail is a mound that appears to be a grave—evidently the final resting place of an individual who died before reaching his—or her—destination. There is another grave-like mound on the Bahn place, too.

The Miles City Trail went toward the east through John Elmore's land where it crossed Chicken Creek—but this is only one of the routes. From the Wahl place there are a number of trails fanning out and then closing in when near Spearfish—many of the area ranches have traces of one on their land—for instance on Billy Evans land, on the Lausser place, on the Redwater Irrigation land, in Sleep's pasture, etc. They are interesting trails to follow even today; however, near Spearfish on plowed land the ruts have disappeared. The two earliest roads to the west were toward Beulah and the other following the Keogh Wagon Road.

At Spearfish the hotel for stage passengers was listed in the local newspaper as Roger's Spearfish Hotel, one of the "largest and best furnished hotels in the Black Hills, with hot and cold baths, etc. . ." so perhaps the Miles City travelers stopped here too. After the stage no longer came to Deadwood, connections could be made with Roger's coaches . . . the hotel advertised a "stage line of coaches leaving the house every morning at

7 a.m. and arriving at 5 p.m." The hotel had been built by Peter Riley in 1877 and then purchased by Rogers.

Spearfish is—and was a lovely town that was started in 1876 when Thomas Jefferson located a ranch at the mouth of Spearfish Canyon, the site embracing several blocks on the original townsite. In May of 1876 a townsite had been laid out and that September the residents built a stockade for protection. By July of 1877 there were one hundred and fifty settlers in the town and a number of ranches and farmers in the area. Spearfish grew slowly but surely and today is a prosperous city—with growing pains. However, in the '80's it was still a village, but promoters of the town urged people to come and see for themselves that the Gem City had made more advancement in the past year than had any other Hills town. Rapid City, they acknowledged, was beautiful but in Spearfish "all that glittered was not nickle-plated" as anyone could see at one glance. With this bit of hometown boosting the paper added that the Spearfish and Miles City Stage had been purchased by J. B. Pruden—this was in 1883.

The earliest road to Deadwood followed the present road somewhat, to the Burton Stockade and then on to Valley Ranche and on to Deadwood following the Sidney Trail. Later the road turned to the south, past the Dryer House on Centennial Prairie—a short distance north of the present Frawley place but this part of the road collected such huge amounts of snow that it was changed to the site of the present road. There also was a road that went considerably south on Centennial Prairie and in 1876 a road went by the way of New Chicago in the Maitland country. However, the stage used the road that went past Dryers as it was mentioned that Reverend Williams had a church near the present Frawley buildings. This church burned in the fire of 1893 and apparently wasn't rebuilt.

Past the present Half-Way House (known as the Cliff House) the early road swerved to the east, past the original Thompson Half-Way House and then joined the present road near the Rest. A short distance from the Rest was the long, steep trail into Deadwood nestled in the bottom of Deadwood Gulch.

In 1880 and 1881 the Hay Creek coal banks were coming to the front and the Miles City mail carrier came that way . . . at least for a time. At this time mail service was once a week from Miles City. There was a road to the coal fields from the Forks on Hay Creek with another route through Beulah. In 1886 the newspaper reported that people coming and going to Hay Creek complained that the old road established in 1876 and traveled since that time—the first "highway" had been fenced across Montana Lake, entailing two more miles of dangerous traveling on a bad road. A frequent complaint in the early days. (Montana Lake is a dry lake bed a few miles east of Beulah.)

The Hay Creek coal fields were thirty-five miles from Deadwood with the early towns being Barrett and Bakerstown. The Homestake Mining Company used Hay Creek coal and everyone thought they would be using it for a long time as estimates were that there was enough coal there to run one thousand mills for one hundred years. In December of 1896 the newspaper mentioned that the only settlement left on the Hay Creek coal lands was Barrett . . . so the early rosy predictions fell a mite short. However, in July of 1898 Nix and associates purchased the fields—1,400 acres of them and soon W. S. Heckert and George Harris were grading a roadbed with their fifty teams. The grade was for the Wyoming and Missouri River Railroad that was being constructed from Belle Fourche, eighteen miles away. The coal beds were extensive, bubbled the local press, and now could be converted to coke by a new process discovered by a Professor Hemingway of Chicago. The railroad would soon be in operation when the thirty bridges on Hay Creek were completed . . . so ended the news item.

The railroad was in existence for a few years—and brought prosperity to the area by shipping out coal, pigs, livestock and hay. But business didn't warrant the operation and the railroad was discontinued . . . and joined the stages and freight wagons in history.

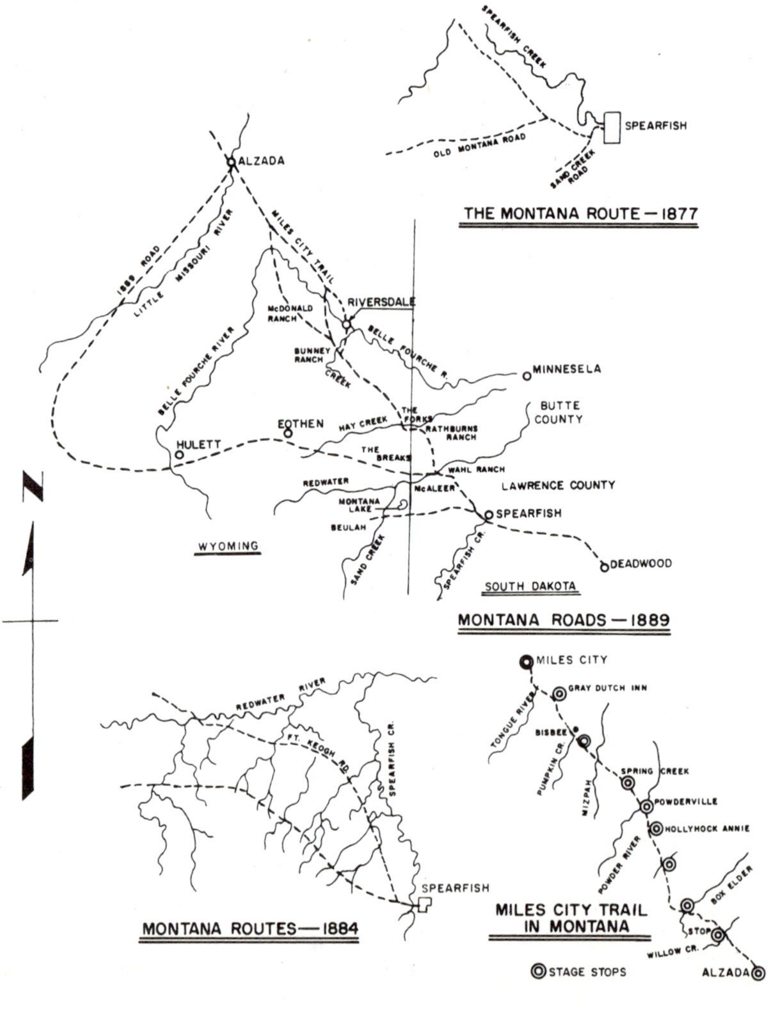

*The Houston Rogers Ranch that was the Gray Dutch Inn when it was a stage stop on the Miles City-Deadwood Route. It is located about sixteen to seventeen miles from Miles City on the south side of Pumpkin Creek.*

*The town of Beebee was located here at the turnoff to the Powderville Road. This was a stable located on the south side of the road; the town was on the other side to the north. The wayfarers stopped here in later days—the regular station was a bit further on.*

*Buildings on the LO Ranch that was a stop in later days . . . the original stage station was toward Beebee on the divide of Pine and High. The LO Ranch appears as a small town . . . houses, barns, sheds and a school.*

*Mrs. Pete Jensen Hill near a building on their ranch. The original ranch was a short distance from here and a stop in later days; however, the original stage station was about four miles further on where Hollyhock Annie operated the place. Nothing is left of the latter stop.*

Pienile, Montana was a 20th century town . . . when the modern roads came in the towns moved to be near the new arteries of commerce or withered away. Pinenile bit the dust in the early '40's.—Photo courtesy Phillis Seaman.

Ruts of the Montana Trail on the Lambert Ranch near Hammond. Pictured are Phillis Lambert Seaman and her brother, Norman, who lived near the old trail in their growing up years.

The Boxelder Crossing of the Miles City Trail. The station was nearby.—Photo courtesy Phillis Seaman.

Alzada, a small town in the southeastern part of Montana. The photo is of an old building at the outskirts of the town. The original town, the early Stoneville, was a bit to the west.—Photo courtesy Phillis Seaman.

North of the McDonald Ranch north of Aladdin, Wyoming was a fork in the road. This went past the old McDonald Ranch to the southeast and to Spearfish. Here can be seen how the rocks were piled to one side of the trail.

*Francis McDonald looking at the site of the old town of Riversdale on the Belle Fourche, that was on the other fork of the road north of the McDonalds. It was a living town in 1889 when the traffic was heavy on the Deadwood-Spearfish-Powderville and Miles City Road.*

*The John Molines on the bank of Oak Creek . . . below the bank was a well for the convenience of travelers. On the far hill can be seen the tracks of the trail. The place was on the original Bunney Ranch.*

*The Forks had a post office located in various homes. This was a stop in later days for travelers. It was near the present Helmer Ranch home.*

*The postman and travelers stopped here at the Rathburn Ranch. It was a short distance east of the Forks.—Photo courtesy Mrs. Alfred Young.*

*The barn on the Rathburn Ranch.—Photo courtesy Mrs. Alfred Young.*

*A stop on the mail route from Belle Fourche to Miles City in later days. The house is a bit west of the present McKay Ranch.*

*An old building on the Bahn Ranch, south of Redwater and not far from Spearfish. This place was settled in the very early days by a man named McAleer.*

*A "hanging tree" near the Ralph Bahn Ranch where the Montana Trail crossed Lake Creek. There was another crossing nearby. The man who ended his life at the end of a rope may have been a horse thief—many of whom faced frontier justice were judged hanging material.*

*Mr. and Mrs. Carl Anderson looking at a mound—apparently an old grave at the side of the old trail. This was south of Redwater.*

*Seth Bullock, a Montanan who came to Deadwood in the early days. He was sheriff, established several stores in the Hills towns and worked for the Forest Service. He also started the town of Belle Fourche and tried to start another town, Englewood, east of Deadwood at the site of Mountain Ranch.—Photo from Black Hills Illustrated.*

*After the Miles City route was changed to go to Belle Fourche, the road went through Colony, Wyoming . . . here is a building of the town on the present James Newland Ranch.*

*Buildings on the Wahl Ranch . . . Mr. Wahl often mentioned the passersby on the Montana Trail who went past his place. Many of them were cowboys, very noisy after an R & R to Spearfish and Deadwood.*

# 19
# BILLY FROST COMES TO THE HILLS

In 1875 William H. Frost of Lamotte, Iowa (twelve miles south of Dubuque), contracted gold fever and to find a cure decided to come to the Black Hills in Dakota Territory where it was reported that gold was to be found in every stream and gulch. He and his party started out on April 22, 1875 and fortunately for us, Billy kept a log of the trip. It took twenty days to cross Iowa, on the 11th of May they left Sioux City and on the 19th were camped on the Elkhorn a bit west of O'Neill, Nebraska, the outpost of civilization at that time. The next day they caught up with Witcher's Train, encamped on the Elkhorn.

May 22nd they had dinner at Little Pine and Uncle Keeler and Myers went hunting . . . the train stopped for the night at Long Pine. On the 23rd a pony express rider came by and informed them that Gordon's Train had been captured and his wagons burned, and that most of the men had been taken prisoner. This was Gordon of the Collins-Russell Expedition—the outfit that had come to the Hills in the fall of 1874—Annie Tallent had been a member of the party too. John Gordon and Eph Witcher went back to Sioux City in February of 1875 with gold they had placered from French Creek—a magic treasure that filled many an individual's heart with a desire to do some placering for himself in the Black Hills—including young Billy Frost.

On the 24th Billy and members of his party had a dance and big time—dances were held frequently so it was almost a necessity to have a fiddle player along—or at least someone who could play a mouth organ, banjo or a jew's harp.

On the 25th a camp was made on the Niobrara after a ten mile journey. Uncle went hunting, got nothing—he wasn't much of a hunter. The next day Billy and George Myers went hunting and had better luck as they got a deer and a goose, but had to spend the night on the prairie. The only thing they had to eat was hardtack, they were afraid to light a fire and almost got devoured by hungry mosquitoes. Frank, their animal—likely a donkey or mule—took off and they had to pack their stuff on their backs. The following day when they got back to camp they were wet and hungry but to lift Billy's spirits there were two letters for him from home.

May 28th was "clair and pleasant." The men tried to find a wolf that Keeler had wounded but with no luck. Others brought in a twenty pound wild turkey . . . they also had come across three mules and two horses on the prairie and brought them to camp.

May 29th they got good news that a treaty would be made to allow the settlers in the Hills. They also had a scare when a gun was fired from across the river and that night double guards were posted. May 30th again was "clair and pleasant" and Billy got busy and washed out a "shirt and hanker."

The next day brought an end to the good weather when it started to rain. They traveled only four miles when they set up the tents. It rained all that night but started clearing up June 1st. One of the men—Old Missoura—went out hunting and saw four Indians. Uncle also went out hunting but wasn't back when Billy wrote in his log.

June 2nd Billy had guard duty from 1:00 a.m. to sunrise. Myers and Uncle Will went out hunting. On the 3rd two men came to camp from Andrews Train and said they expected to go back with Witcher. June 4th was "clair and cold." Pat went out and killed two antelope. The following day hunting was poor—Uncle hunted all day and got nothing—but this wasn't too surprising because a Daniel Boone poor Uncle wasn't. June 6th was a cloudy and cold Sunday. The party had fish for dinner and Doctor brought in a black tail deer. The next day they traveled six miles. June 8th it rained until noon—the 9th the train went eighteen miles, a considerable distance. Billy got a letter and learned that a treaty with the Indians wasn't imminent and that there were 30,000 Indians on the war path. Billy wrote in his log that Witcher had gotten his back up that day and they were having a meeting that evening to elect a new captain. The following day seven wagons joined the train. June 11th they traveled fifteen miles and camped on Rapid Creek. Men in seven of the wagons decided they'd had enough and started back for home. The next day they again traveled a good distance, sixteen miles. On the 13th they had dinner two miles west of the Snake River and in the afternoon camped on the Niobrara. Poor Frank was ailing. The next day Witcher drew a revolver on Robert and then hit him on the head for good measure. The party got a bonanza—three antelope and six deer—and more good news, Frank was feeling better. The next camping place was on a high hill on the Niobrara—that evening they had a dance.

June 16th the train traveled eight miles in warm and sultry weather. June 17th they came on the spot where Gordon's train had been captured . . . "the place looks like the ruins of a warehouse after a fire, wagons or parts lay around in wild confusion. Also, guns, ammunition, shovels, picks and pans, in fact everything that is used in the mines," wrote Billy, then added, "It indeed looks hard. In the center of the plat is a cottonwood pole erected by Gordon himself and on the pole a board with this inscription. 'Friday, May 21st Gordon Party captured in this place by Capt. Mills of Fort Laramie. Capt. Mills is a d_______s_______a_______.' "

The party stayed here until the 20th. The 18th it had rained all night but cleared up on the afternoon of the second day. Uncle went hunting—with the old refrain—he got nothing. Three of the men went scouting and met about one thousand Indians—one man got away but the others had their guns and blankets taken. On the 20th they moved about one mile and on the 21st were visited by two hundred Sioux who were on the warpath to fight the Pawnees. The 22nd of June the party traveled about fifteen miles and crossed the Niobrara. The party was planning to split up—some going to the Hills, the transportation company to Spotted Tail and Billy's party to Fort Laramie.

For several days their group went toward Fort Laramie hoping that the Hills would be opened. Uncle went fishing and got a large string of fish—he was a better fisherman than a hunter. On the 4th of July, a cold and stormy day, they passed the ruins of a mail post that had been burned by two white men. Billy and party crossed the Running Water in the morning and from here expected to make a break for the Hills. They came about twenty miles that day, over endless sand hills. On July 5th the train traveled fifteen miles in the morning to Rawhide Creek and eight miles in the afternoon when they saw government freight wagons. Men in the wagons asked where they were headed and the spokesman said they were hunters after some game. If they had said they were going to the Hills they would have been placed under arrest.

July 9th they camped on Old Woman's Fork after traveling about twenty miles. The next day they traveled through some Badlands and on the 11th came to the south fork of the Cheyenne, where they stayed awhile on account of high water. July 12th they left camp and crossed the south fork of the river four times and that night had a dry camp. The 13th they arrived at the foot of the Hills and the next day came to Camp Jennings—Billy meant the Jenney Stockade near the present Newcastle. On the 16th they arrived in Custer, left the next day for Spring Creek where they staked out some claims—which didn't prove to be productive. Billy wrote of working the claim, of deer hunting, cleaning and about the wagons passing by. Then for a time there were no entries.

In August Billy started working for H. B. Kelly as a ranch hand—this was near Cheyenne where he worked until February of 1876. In later years the young man was in charge of a LaPlant bull team on the Fort Pierre Trail and after his freighting days were over he settled in Pierre.

William's journal gives a good account of the days when the Black Hills were the Gold Hills—forbidden but tempting—and of the men who slipped in to share—hopefully—the golden treasure in the streams and gulches.

# 20
# AN ENGLISHMAN VISITS THE BLACK HILLS

Edwin A. Curley came to the United States from England as a correspondent for the London newspaper, "The Field." Curley's articles included information about Nebraska's potential and future for coming settlers and he also compiled a book, "Nebraska, Its Resources, Advantages and Drawbacks." The Black Hills area was coming to the front and Curley decided to make a trip there himself—not for seeking gold but to gather material for a guidebook; "Glittering Gold: The True Story of the Black Hills," that was published in Chicago in September of 1876. The following year Curley returned to the Hills and gathered statistics on the population, cost of living, the various routes, agriculture, etc., that appeared in a second edition published in June of that year and entitled, "Curley's Guide to the Black Hills." The original cost of the book was fifty cents!

In describing Cheyenne, Curley wrote that "Few towns of its size transact an equal amount of business." He checked on prices in several stores and ascertained that competition was lively and supplies for the pilgrims were comparatively cheap. He maintained that merchants freighting by the carload could afford to sell articles cheaper than they could be purchased in the east. The Cheyenne people had large amounts of money invested in cattle, sheep and horses, there were many stock ranches in the area. The town had quite a few celebrities, some of them having gained wealth by "pandering to the prevailing dives of the border town."

The Englishman and his artist companion booked passage with the Cheyenne-Deadwood Stage and "took places on the upper deck," to Fort Laramie, ninety-three miles away. The road to Pole Creek, eighteen miles away, was in top shape. Horses were changed but due to the scarcity of fuel dinner was eaten ten miles further at Horse Creek where there was plenty of fuel as well as water and grass.

Fuel was a problem on the prairie—Bear Springs ten miles down the road had very little wood. In fifteen miles the Chugwater Valley was entered where there was fuel, water and grass . . . and grandeur and beauty in the lofty "sand-rocks carved into fantastic shapes or fortifications, or semi-architectural forms, by the light, slowly acting yet all-powerful forces of nature; the verdant valley a smooth meadow of rich grasses and wild flowers; the pleasant stream bordered with trees and with bushes of plum, cherry, and currant, and willow and alder; while lowing kine and playful calves around, and the camp-fires of freighters and pilgrims give life and animation to the whole." The stage stopped for the night at Phillip's Ranch.

The second day's ride was mainly along the Chugwater, they passed Chimney Rock. Hunton's Ranch was passed. James Hunton, according to Curley, had been murdered and scalped a few days before and his horses stolen. Eagle's Nest was approached which was an opening to the road of a narrow deep ravine—a good place for an Indian attack. The stage arrived at Fort Laramie early in the afternoon—and from here the travelers had to make other arrangements to get to the Hills. Laramie City was described as a pretty place at an elevation of 7,123 feet, many gardens and young shade trees—second only to Cheyenne as an important city in the territory.

Curley was able to get a buggy and two horses and they were soon on the way to the new El Dorado. They passed the Rolling Mills, past the ranch of Dutch Molly, the milkmaid, four miles out and up a divide. There were occasional patches of snow and fields of flowers were so thick "as to render their rich purple color visible several miles away." Ragged Top, a "pinnacle of rock" that "offers many beautiful beautiful scenes to the lover of the picturesque," was passed. From the mountain a descent was made for several miles to Horsecreek where Dalton's Ranch was located. Dalton, a fellow Englishman, was well-to-do and the "interior and surroundings of his establishment show more evidences of taste and refinement" than Curley had seen under similar circumstances.

From Daltons was undulating prairie to Bear Springs, fifteen miles distant. (Plans were underfoot at the time to make a shortcut road along the Chug—eventually it was done and became Dalton's Cutoff.)

The traveling party with whom Curley and artist traveled was a "heterogeneous" one, with freighters from several states, pilgrims who paid for their baggage to be toted but walked themselves, there were some merchants and a few unattached people on horseback . . . the party soon added up to eighty-nine people, most of them well armed and thirty wagons. Curley's group organized, a captain and a wagon master were appointed and watches were set up. The group was at the Government Farm at the time.

It was a Friday morning when the party left, crossed the Rawhide and passed Rawhide Buttes, a "grand mountainous peak of eruptive or metamorphosed rocks." That night a camp was made on the creek where there was plenty of good wood and water. Next morning they went up a somewhat difficult hill and followed a high plain for some miles and then went through country that would be advantageous for an Indian attack. That evening camp was made at Running Water . . . with plenty of good grass and water—but little wood.

Curley was not pleased with the wagon master and labeled him very inefficient as he did not send scouts to look over the trail. Several Indians were observed in the distance and when they reached Hat Creek (Curley believed it was Old Woman's Fork) they had a skirmish with the red men. Curley described a few specks appearing over a rise of ground and the cry went up to secure the horses. Shortly all horses were corralled—some inside the corral and others inside a circle of riflemen. The Indians went past as a whirlwind and shots were fired but that was about the extent of it. That night rifle pits were dug as Indian signal fires were seen at several points but there was no attack.

The party started out for Indian Creek, so called by Black Hillers but it was called Sage Creek on government maps. This was "undulating country, easy to traverse, excepting that there are numerous deep, dry ravines, with almost perpendicular banks, which in the absence of bridges or culverts compelled us to go much out of our course." These ravines offered many excellent hiding places for the Indians and for the road agents too but on this day they were scouted and proved clear. The group camped on Last Water on Indian Creek, dry at this time of the year, but some dug for water which proved to be of poor quality. Again guards were set out and the night proved to be uneventful.

The next morning there was discord in the camp and the people separated into two parties . . . but camped close together on the Cheyenne. Nearby were graves of people slain by the Indians, but this evening Curley's party was unmolested. The next morning they started out and at the mouth of Red Canyon found this inscription on a tree: "Look to your rifles well, For this is the canon of hell; the Red Canyon." And indeed this was a dangerous canyon for attacks within the narrow sides and lofty sides "probably more than a thousand feet high," and with many ravines leading off to the sides. The canyon was scouted and one party thought they saw some Indians but Curley and others were "prone to think that no Indians were seen by the others."

During the day Red Earth Creek was crossed many times, a miry stream that gave them considerable trouble—however, the scenery was beautiful. Pleasant Valley was entered, with the "butes and bluffs being banded and spotted with red earths and white porphyritic limestone, and sometimes with yellowish and greyish shales, which with the corresponding stripes of light green grass and dark spots of pine, made curious and not unpleasant landscapes, which it would be difficult to render by the engraver's art in any but a technical way, unless with the use of chromatic coloring." The party stopped for their noon day halt about a mile apart. A body of a man and three dead horses were seen. That night the parties camped close together—the dead man and animals evidently worried them

—but they were safe. They set out at four in the morning for Springs on the Right about four miles away where there were good springs. They arrived during the day in Custer.

"Custer," wrote Curley, "is very prettily situated in the spring of French Creek, with islands of pine-covered rocks springing from the grassy meadow all around it and beyond these islands on the left, in the direction of Harney, mountains of considerable magnitude. The crooked creek lay hidden in the very bottom of its bed sleeping so lazily that it could scarcely glisten in the sunlight."

Log huts and houses were in abundance—but the owners had vanished up Deadwood-way that was the current magnet of the gold seekers. The mayor, whom Curley visited, estimated that there were four hundred people still in the town and about eight hundred homes, although Curley counted them and got the total of 501 buildings and about two hundred inhabitants. There was very little mining being done on French Creek for want of enough water for sluicing—a common complaint of the early Hills miners. Agriculturally and pastorally, Curley thought the region very good . . . and he also estimated that the gold region here would last much longer than Deadwood. Well, a man can't be right all the time!

Leaving Custer they passed the Santa Cruz Peaks—so called because there appeared to be a huge Latin cross on one of the peaks. The foreground was beautiful and the scene one of the finest to be found anywhere. About eighteen miles from Custer was Hill City where there were some one hundred and fifty homes—but no inhabitants as they, too, had gone to Deadwood Gulch. This also was beautiful country and Curley's artist made several sketches . . . including two views of Spring Creek that they crossed twenty-three times on the way to Deadwood. They passed Golden City Park where the diggings were located from which Professor Jenney had formed his opinion of the Spring Creek diggings—nearby was the outcropping of the Mammoth Jenney Lode.

Between Spring Creek and Rapid on the Deadwood road were ten miles of quite rough road . . . but interesting for the agriculturist and herdsman. There were also a number of petrified trees to interest a geologist. At the Crossing of the Rapid was a log cabin of the office of the Recorder of Mines. The Rapid was described by Curley as the largest stream in the eastern part of the Hills with several fine tributaries with much gold in the main gulch. The Box Elder was a fine stream too with numerous pretty valleys and small bodies of good land where a number of people had settled. Some had fenced in hay pastures and others had planted potatoes . . . which Curley thought would make a tidy profit for the farmers. Elk Creek was crossed next with Elk Mountain (Custer Peak) close by. After crossing the creek they ascended the Long Hill which was very steep—the worst Curley had ever seen. He carried a pail of water from the creek for half a mile or more to their camping ground—and then discovered water in a ravine near their camp. The greater part of the trail was very bad—with very little grass at the sides due to a fire about eight years previously, making the area dreary and dismal; however, there were some gorges where there were "fine cones of spruce, intermingling with the pines and patches of deciduous underbrush among the rocks gave a very pretty effect." This road was miry and there were big boulders but after crossing Bear Butte Creek they came on a turnpike road that was a decided improvement. Whitewood Gulch was reached and Curley mentioned Whitewood City, a little cluster of log houses in a narrow gorge about one mile and a half up the stream and then Deadwood City, a larger camp, half a mile further. Deadwood was filled with miners busy working sluice boxes . . . all trying to make a million dollars at least.

Curley visited Crook City, mentioned the other cities in the area and then went back to civilization by the way of Sidney.

# 21
# W. H. COMES TO THE BLACK HILLS (Will Hamilton)

In December of 1883 W. H. (Will) Hamilton left West Virginia for the west and worked for a time in Iowa and Nebraska. In December of 1884 he got a bad case of wanderlust and on the 16th started out for Deadwood, Dakota Territory. He went by train as far as Sidney and then boarded the stage (the company honored his railroad ticket) a lumbering vehicle. Departure time was 9:00 in the morning of the 19th with the temperature thirty-five degrees below zero and with the snow drifting. It wasn't long before the windows were frosted over so the passengers didn't see much of the country except when the coach got "hung in a snow drift" and the passengers got out to help lighten the load. Sometimes young Will helped shovel, but not one of the passengers, a "soldier boy" who was returning to Fort Robinson from a furlough in Missouri. The superintendent, J. M. Strawther, was riding on top with the driver and he wasn't a bit bashful about wielding a shovel.

The stations were ten to twenty miles apart, depending where there was water and land to make hay for the stock. At each station was a stock-tender—most of the stations were a fourteen by sixteen foot one-room shack, usually of logs with a ceiling just high enough to allow a tall man to stand. The roofs were of boards covered with six inches of dirt, which made the buildings cooler in the summer and warmer in the winter. Oat sacks were kept in the houses which came in handy as beds in emergencies.

Meals were good—plenty of meat, potatoes and baking powder bread with the cost from six bits to one dollar. Will thought the station tenders made a special effort to have good meals due to Mr. Strawther being along.

The North Platte was crossed about midnight on the 20th and the coach arrived at the next station just after noon. The wind blew fiercely and

Mr. Strawther and the driver became completely exhausted in the thirty below weather and decided to spend the night at the station. There were plenty of buffalo robes, grain sacks, blankets—and wood for the fire. Will wrote that he spent a fairly comfortable night but one of the passengers, an elderly woman named Mrs. Miller said she was more than fairly comfortable—she slept as a baby. The wind died down during the night and the travelers started out at nine in the morning with fresh horses. But they didn't stay fresh very long as the snow was packed, ". . . so it made terrible traveling and I am sure we did not make more than two miles per hour, having to shovel in all the low places and getting the team down many times," remembered Will many years later.

It was almost night when they arrived at the next station and everyone was famished. Will said he certainly ate his seventy-five cents worth that night. The driver said he was beat and couldn't go any further so they spent the night. Most of them had badly frost-bitten ears, noses and fingers and all complained of being sore. It was rough country in which they traveled the next day and they kept on the high country following the ridges but there was shoveling to do in the creek crossings and ravines. They arrived at the next station a little after noon, ate, changed horses and went on after dark.

The next station was Pumpkin Creek where they had a snack and started out again about midnight. They had a bit of a delay when a wheel got stuck in an old well but they got logs and poles and lifted the stage and were soon on the trail. They arrived at Fort Robinson about 10:00 a.m. where they had dinner at an eating house. The soldier and Mr. Strawther stayed here. The coach was traded for a pair of bobsleds with a wagon box on top, driven by a new driver—a green man who had been on this part of the route only once before.

The coach left the fort at 3:00 p.m. with the road visible only in places. When it started to get dark the driver suggested they stay overnight at a nearby ranch where they had a good supper, good beds and W. H. had a chance to remove his clothes for the first time since leaving Sidney.

The next morning was Christmas when they started out about seven in the morning—it was clear and cold—but no wind and they soon arrived at Big Cottonwood ten miles away. The sled had made better time than the coach and they arrived here about half past nine in the morning and were on the road in one-half hour after warming up their chilled bones and changing horses. The stocktender said he had ordered groceries but the stage hadn't delivered them and he had on hand only potatoes. The coach was on the road about one hour when it began to snow and blow and although it was fairly level country they had to shovel in the low places. At sundown the wind died down, they saw a light in the distance—and to their surprise they were again at Big Cottonwood—they had traveled in a giant circle.

The stocktender boiled up a big pot of potatoes—and W. H. thought they tasted real good, too. They slept on sacked oats and with buffalo robes for blankets got through the night fairly comfortably. They started out in the morning and soon a hard wind came up and traveling was as difficult as it had been the previous day. One time when Will was helping shovel snow at a dry creek a rabbit jumped out of a clump of bushes and he shot it with his "wonderful .32 six gun." This shot frightened the horses and they took to their heels—with Mrs. Miller aboard. The driver said they wouldn't go far—which they didn't. As Will described it, "The leaders got down and the hindmost team piling in on them in a bunch." Mrs. Miller, described by Will, was a wonderful woman, cheerful, full of jokes, uncomplaining . . . a true frontier lady. She was unaware of the runaway as she was completely covered with blankets.

The wind went down again, they saw lights in the distance and whipped up the horses to get to the station—and lo and behold—it was Big Cottonwood again! They had traveled two days in a circle!

There was meat for supper—the jackrabbit—although it was tough as plow boot leather; however, the broth flavored the potatoes. W. H. kept a fire going all night and the rabbit was pretty well done by morning. The weather had moderated a bit but Will announced he was going to wait until the weather got a lot better—or they "ate all that boy's potatoes." The driver decided he would wait for the next stage to come through and which arrived about four in the afternoon. The stage—a sled—had three passengers including Tom McMaster and Mac the Saddler.

The wind did not blow all day and since it was a moonlight night the driver said he would follow the backtrack to the next station, Willow Grove, seventeen miles away. "The track was plain and the night was bright and still, and I believe one could hear that sled squeak for two miles," commented Will. They got to Willow Grove about midnight. Here was plenty of food—hot bread, steak, fried potatoes and coffee and W. H. commented that he had never seen so much grub stowed away by three people in so short a time.

The passengers slept until eight in the morning and started on before noon. It was cold but not stormy; however, they still had to shovel in the low places, crossing ravines and the creeks. They traded their bobsled for the coach before getting to the Hills, taking the coach left by the southbound party. The streams were open in approaching the Hills and at one creek Mrs. Miller took a tin cup, reached out and got her fellow passengers a drink of water that was almost milk-warm.

The exhausted and frost-bitten passengers were happy to arrive in Rapid City on the evening of the 29th. The driver said he wasn't going to Deadwood that night—even if he lost his job. W. H. stayed in a hotel and relished the good bed and a chance to take off his clothes for the second time since leaving Sidney.

The next morning was a chilly forty below zero when they set out once again about eight in the morning. They skirted the Hills, keeping on the east and north side. Horses were changed at Blackhawk and Sturgis and they came to Deadwood on the Crook City route. It was six in the evening when they arrived in the gulch city.

Will Hamilton stayed in Deadwood for a few days, then worked at Minnesela for a time and settled on a farm near the present Belle Fourche. After several years he sold his place—this was in 1901—and went to Missouri.

# 22
# F. M. LEAVES THE BLACK HILLS
## (F. M. Allen)

The early Black Hillers liked to use their initials instead of their given names—and Mr. Allen was no exception. He left the Hills in 1883—perhaps only on a visit—by boarding the Sidney coach at Sturgis on Wednesday the 31st of January. There were three other individuals already aboard—Porter Warner, editor of the Black Hills Daily Times, Paul Gentges and Frank Karper. The weather was pleasant and the roads were good and they arrived at Slate Springs the following morning where they had breakfast.

They were on the road only a short time when it began to snow but the roads remained good and they arrived at Lone Wells at nine o'clock, where they transferred to an open wagon box on runners. Willow Grove, the next station, was reached about noon where they had a lunch and the horses were given oats. This was their last meal for fifty-six hours.

They arrived at Big Cottonwood and although the snow was drifting, the driver by the name of "Red" thought he could make it before the tracks were completely obliterated. It got colder and colder, the wind blew a gale and they could not see ten feet in front of them. The horses floundered in the drifts and in extricating them one time they discovered they were lost. By this time they were in the Nebraska Badlands where there were ravines and washouts—there was nothing to do but stay put for the night. The horses were unhooked and tied to the sleigh. There was an old worn-out robe and Mr. Allen had some California blankets to cover the passengers but to keep warm they had to get up frequently and dance a hornpipe and flail their arms to keep up circulation. In the morning the sun came out briefly and they got their bearings.

Frank Karper walked in front to lead the way—and became lost . . . the passengers and Red were sure they would never see him again. At noon the sun came out for a moment and they again got their bearings—they had been going to the west instead of south. They stopped at a dry creek where there was cottonwood and dry brush with which they built a fire. They turned the wagon box on its side for a bit of protection. Four or five hours later Frank showed up almost completely exhausted but revived with some whiskey. In his valise he had a small link of sausage and three small biscuits which they gobbled up in no time flat. They later learned the temperature registered forty-two degrees below zero.

The morning was bright and cold. Karper and the driver each took a horse and went off in search of Big Cottonwood. The three others kept hunting for switches to keep the fire going. It was almost night when help reached them—fresh horses, oats for the two remaining horses and food for the men. F. M. said, "Three feeble cheers went up from our camp for the deliverance at hand. We gathered up our traps and started out, reaching Big Cottonwood after dark, and then going on to Red Cloud."

This account, as well as the other stories of early travelers, tells us that taking a trip in the early days was for the hardy, especially when winter ruled the prairies.

# 23
# THE TOLL ROAD TO DEADWOOD

At one time all the roads that came into Deadwood were toll roads. Sometimes they were in excellent shape but more often they were in terrible condition. One traveler called them a luxury; a conglomeration of rocks, mud holes, logs, broken bridges and loose courdoroys, available to anyone for twenty-five cents. Not to have traveled these trails was to have "missed all the thrills and estatic delights in human existence . . . then wait twenty-five minutes in front of a toll house to await the pleasure of shock-headed, grease-besmeared, overalled individual who works the pole by courtesy called a gate, in order to pay him the privilege of riding over his man-trap. After biting the change and giving a few growls supposed to be words, he then opens the way and allows you to pass to your doom." And the wayfarer concluded that this was one of the varieties of western life.

These greatly maligned roads played an important part in the transportation history of early Deadwood as the long distance traffic, as well as the local travelers used them. The earliest comers to Deadwood Gulch selected the steepest places to get down to the camp. The gulches were chocked with fallen trees and brush and for this reason the travelers used the ridges—also there was less snow on the high places in the winter months. Buckskin Johnnie—J. T. Spaulding—told of entering the gulch city by unhitching the teams of horses and with two men at the tongues of the wagons lowering them from tree to tree. Later trails were made into Boulder Canyon where it would be necessary to roughlock the wheels on the steep inclines, that is, fixing them with chains so they would drag instead of turn. An early freighter said it took the best part of the day to lower the twenty wagons of his outfit.

One old-timer, with a perfectly straight face, said the roads were so steep that the horses would grasp the tree trunks with their teeth to ease the way down!

The earliest roads to Deadwood City were from the north and from the south—the latter the Cheyenne-Deadwood and the Custer Road. From the north came two roads from Crook City and one from Spearfish via New Chicago—the later Garden City and still later Maitland.

Edwin Curley told of the Cheyenne-Deadwood Toll Road and blessed the man who constructed it. Early owners of the road were Philbrook and Wannemaker who later sold it to Cummings and Cline who put a large number of men to work on it. A bridge was constructed over Whitewood and the owners felt they could then charge twenty-five cents for a pack animal and one dollar for a single team. One dollar and fifty cents was charged for a wagon with four animals without getting too much static from the traveling public who were known to complain mightily if the roads weren't in good condition. And usually with considerable justification.

The Cheyenne Stage used this road until 1877 and the Custer Stage Line for a longer period of time. Martin Brosnahan, an early settler in the Hills, said the Custer stage went past the later Monarch Mine in Upper Two Bit Gulch. Later the Custer Road was over Lexington Hill, past the Lexington Mine, Spruce Gulch and entered Whitewood Gulch near the present rodeo grounds. There was a toll gate at the mouth of Spruce Gulch and in early 1878 a party of men working in the sawmills up the gulch attempted to make a new road around the gate . . . they said the road was out of repair and the rates exorbitant. The gate keeper saw them, came out and had them desist—with the help of a gun

By 1882 the Custer stage was no longer using the Spruce Gulch route—at that time it went by the way of Central City from the Eldorado Hotel, to Centennial Prairie, Spring Valley and Battle Creek and by 1884 it was going on the Cheyenne Stage Road to Ten Mile.

Several individuals and companies attempted to make toll roads up Deadwood Gulch to Gayville and Golden Gate City—one of the first mentioned was in August of 1876 when Kline and Company spent two thousand dollars on a project. The newspaper complimented them by saying this road would be a decided improvement over the old road that went over the "big hill that had been a terrible deal."

By the following year there was a storm of complaints about the gulch road and some prominent citizens talked of constructing their own road, putting Deadwood Gulch on the right. They even got a right-of-way from the miners who pledged not to dig up the road with their placering provided the piling was set on bedrock. Miners frequently dug up a road if it interfered with their operations and one time when a road company brought one to court, the judge ruled for the miner, who, after all, was working his own ground.

In 1878 road conditions hadn't improved and accidents were almost a daily occurrence. Foolhardy pedestrians tried to see how near they could get to a team of horses without getting injured and this didn't set well with the teamsters on the very narrow road. They complained bitterly stating that there was barely room for a wagon between the gulch and China bagnois—South Bend, especially, had a large Chinese population. The public was hopeful that the road would be better after J. J. Bump, the original locator of the Deadwood-Central City Toll Road bonded it to C. C. Fyler, who turned out to be quite a rascal. Old C. C. would not pay for it and Bump wanted it back, but then Fyler decided he would pay for it after all . . . and even made several improvements on it. A fountain was put for the travelers and a watering trough for the horses and other animals. At one time Fyler purchased one hundred and five baby chicks, which were quite an attraction for the travelers passing through. In May of 1878 it was noted that the road was in perfect shape.

In January of 1879 the toll house and outbuildings burned to the ground; however, Fyler got busy and immediately rebuilt them. He also established a livery and hack line that he had to sell when he got in some financial difficulties and his property was sold at a mortgage sale—all except two horses that he used in making his get-away.

A Mr. Jabez got control of the road and hired S. A. Crist as gate keeper—Mr. Crist also had charge of the county hospital across the way. Mr. Crist evidently was a good, steady man as he had the job for many years. He planted a beautiful flower garden near the gate that was much admired. At times he would have the snow shoveled from the road to the sides, making it easier for the traveling public.

Two gentlemen who remember passing through the toll gate as young boys are George Glover of rural Lead and George Hunter of Deadwood. Mr. Glover remembers that one of the Crist boys gave him a German Shepherd puppy but his father thought the seller should be reimbursed and gave the boy one dollar. George had the dog for fourteen years until it went blind.

In March of 1899 there was a fire on the roof of the old toll house and people rushed to the scene and threw snow on the blaze and in one hour the fire was extinguished. The old frame structure with a connecting hay barn was owned by C. A. Scott of Spearfish and occupied by John Landried and family. Landried was a section foreman on the Elkhorn Railroad that went up the gulch and then up the hill toward the Washington section of Lead.

There are faint traces of the toll house, down from the county highway yard and approximately opposite the present Bob Ruth residence.

In 1877 Otto P. Th. Grantz obtained mining claims that extended the full width of the gulch in Deadwood, near the present Sunnyside Motel. In 1879 a road was established to Central City that went through the property and Grantz didn't mind in the least until in 1887 when the city fathers planned to extend Main Street. They offered Grantz five hundred dollars which he refused . . . and to back up his refusal erected a fence to block traffic. The authorities tore down the fence that the mining man immediately rebuilt. This happened several times and finally Grantz was arrested for obstructing a highway—but this didn't bother him at all and he erected the fence again. The city fathers gave up for a time, they fixed the Williams

Street Road but the people complained that it was dangerous and narrow with only two places where vehicles could pass each other—this was the stretch of road between the upper end of Williams and Mrs. Livingston's house. (A bit further to the west, near the present Dr. Smiley's home there was a toll gate at one time as shown in an early photograph.) In 1897 the city again made preparations to open the lower road and the intrepid Grantz put up a sign proclaiming, "This is not a public road. It is closed." Eventually the city and Grantz came to terms and the road was once again opened to through traffic.

A road from Deadwood Gulch went to the north up Blacktail Gulch, to Garden City and Spearfish. This road called the Blacktail-False Bottom Toll Road was owned by Richard Griffith with the toll gate the present Meredith Nelson home—fixed up since the early days. In 1888 the toll road was destroyed by a flood but was shortly repaired.

When the Ragged Top mines came to the front a toll road was built on the north fork of Deadwood down to Squaw Creek by John de Launcey. By the middle of October 1889 two miles were already completed.

An early road to Centennial City was built in July of 1876 by a company composed of W. L. Kuykendall, E. B. Farnum and others. The project cost $7,500 but everyone thought it well worth the cost to connect the two cities—although Curley believed Centennial City was probably not destined to become anything more than "a fools-cap town." He described the place as being some eight miles northeast of Deadwood, in the open prairie, with beautiful hills around it. It remained a town for a number of years with homes, a saloon and a ten-room hotel.

The Centennial Hotel, at the site of the present Henry Frawley Ranch, was a bit to the west of Centennial City, and according to Mr. Frawley, Sr. the place was known as Soysters when it was a distributing point for the mail going to Spearfish and Minnesela. The Reverend George Williams was the postmaster; he also had a church where he held services until the building burned in September of 1893. The post office was discontinued in April of 1897.

This fire of 1893 was a bad one that burned houses, barns, haystacks, sheds, the Centennial Hotel, it went up to the Charles Pfrunder Ranch and burned all the buildings there as it did at the Stearns Settlement about one mile from the hotel. The family at the settlement fled to the lime kiln, on the present road to Deadwood, but two large piles of wood caught fire and made the place an oven. The family members fled to the Cliff House that in later years was known as the Half-Way House, which also burned. They went on to Deadwood where they received medical attention.

The fire went its hungry way to the original Half-Way House, operated by Mr. and Mrs. Pete Thompson but didn't get the buildings there which included a large roadhouse and barn. (Nothing is left of the place now except some weeds . . . even the spring across the road has ceased running.) The fire in its erratic course turned to the east and threatened the section house at the mouth of the railroad tunnel and as a precaution all the furniture was removed from the building. Fortunately the wind again changed course and the house was saved.

This was one of the biggest fires in the Hills—it burned areas near Lead, Portland, Garden City, Two Bit and toward the Bulldog Ranch. It was said that people ten miles from the raging inferno found the heat unbearable and places miles and miles away had pieces of burning bark land there. Even in Pierre the flames could be seen painting the night sky. Above the Carstenson Ranch are the remains of a little stone cabin with the wooden parts burned away—no doubt, in this fire.

A new route was made to Centennial City and Crook City later in 1876 that left Whitewood Gulch a short distance from Deadwood, went up Deadwood Hill, up a gulch where the present-day Terrace Motel is located, past a gravel pit, near the later Deadwood Dick's cabin, through the site of the Pinecrest Park, to the head of Slaughter House Gulch and through George Keller's Ranch—the later Spindler Place. It then went a wee bit east of the present Preacher Smith monument, the area known as The Rest in the early days. The trail apparently followed rather closely the Centennial Wagon road to the prairies. This was the Nobel Toll Road that gave access to the agricultural district from the north and northwest.

In 1894 the Deadwood City Council was debating whether they should buy the road or not—and the decision was to do so. By March of the following year people were complaining that it was in worse shape than it had ever been before although William Leemaster of Centennial was paid $250.00 a year to keep it in shape.

The Nobel Toll Gate was located at 308 Main. In 1928 the building was moved to 300 Main when it became part of the Sinclair Gas Station built by Bill Auer who operated the station until 1964.

In 1880 a road was established known as the Whitewood-Deadwood Toll Road that followed the valley of Whitewood from Fountain City to Raspberry Gulch and not far from this side of Whitewood Falls. It then went up that gulch to near the divide and from there over the ridge was an easy grade to Centennial Prairie. The people figured it was only four miles to the prairie from lower Deadwood and the route would be shortened by three miles. Only one gate would need to be established . . . this might have been near the present slag dump on Whitewood.

As can be seen, the roads changed rather frequently—part of the road would be new and then join one of the old roads. Or the old road would be used part of the way and then strike out in new country. Many a time a road would be established, fall into disuse and then be renovated—for instance the Old Blake Road—the lower Crook City Toll Road that was reopened in 1878. It was noted that it was in superb condition with the bridges all repaired and the entire distance planked over—good news for the workers in the seven sawmills in the area. Courdoroyed roads were good in wet weather but made for mighty rough traveling by wagon—or even by stage and buggy.

This road left Crook City in the southwest corner of the settlement and worked southerly—with numerous curves—through the Grenfell place, over Best Hill on Albert Best's land, crossed Whitewood Creek, Sandy Creek where nearby is Shannon Spring that is Whitewood's water supply. It then went south of Crook Mountain and toward Boulder Canyón where it branched. One trail went past the John Anderson Ranch, where the ruts can still be seen, through the Oak Ridge Cemetery, past the race track, crossed the present road and entered Deadwood by the way of Split Tail Gulch. In lower Deadwood Whitewood Creek took a different course than it does now—it was quite a bit to the south and the early road went on the valley side opposite of the present road.

One of the first roads declared a free road by the county commissioners was the Oak Flats Toll Road, halfway between Deadwood and Crook City when the commissioners gave orders that no tolls be collected thereon. It was fixed up and in March most of the freight coming to the Hills entered by this road—for one thing there was more grass here for the animals than on the Boulder Canyon Road that was used by the stages.

Crook City at the other end of many of these early roads was one of the earliest camps in the Hills. There was an old and a new Crook City, the first about one-half mile from the later camp; it lasted for only a short time. In the last few days of May 1876 the new Crook City had been in existence about eight or nine days according to Curley and had a dozen log houses nearly completed and as many more started, "while tents abounded and the total population may have been two hundred or more." Crook City, as most of the early towns, prospered for a time, then waned, then prospered again. It was first built up with placering on Whitewood and as a trading center, it became a farming community although some placering was carried on throughout the years. When the railroad came to the area, Whitewood was born a short distance to the north and Crook City dwindled to a few log buildings. Today it has several homes along the old street with one of the old buildings remaining—the present Tomovick home owned by the Leemasters in previous years.

In August of 1876 Preacher Smith was on his way to Crook City when he was killed . . . the location given in the early papers as six miles from Deadwood and two miles northeast of the Rest. The Methodist sky pilot was on his way to deliver a sermon at Crook when he met his death—at the time people said Indians were responsible and even claimed that the brave who committed the dark deed was Little Big Man, an Ogallala Sioux. Other people maintain it was a white man or men who killed Preacher Smith, perhaps some saloon keepers who didn't see eye-to-eye with the man of the cloth regarding drinking—and there was a saloon a short distance away that was in business in 1876. Many of the ranches on the trails sported a sign in front of "their dingy establishments proclaiming that it was a 'saloon'," according to an early traveler. Money was a scarce item and early settlers were happy to get a bit of cash from the pilgrims selling them a dram or two or more of select lightning, a meal or perhaps some hay for their animals. This particular saloon was located on the present site of the Duane Carstenson ranch and is a part of the old home.

Later there were other public houses including one owned by Charles Zimmach who had one five miles out of Deadwood where he offered free samples of buttermilk to the traveling public. The Junction House was located on the Spearfish and Whitewood roads intersection—a short distance from the Cliff House. The manager was Jack Anchor.

In 1889 Amos J. Cummings took the road from Sturgis to Deadwood via Crook City that was full of "antiquated contrivances from the days of Queen Anne with white covers and a boot 'that do paralyze Delancy Kane,' usually overburdened with baggage. One hack, a three-seated rockaway, carrying a Titanic broke and took three hours to get to Deadwood." He described the Crook City hamlet as a collection of deserted log cabins and frame houses although it once had 1,700 people.

On the road to Deadwood was a string of Conestoga wagons loaded with supplies, frequently one-half dozen of them strung together with chains and with two to twelve spans of mules and steers attached to a leading wagon. The driver was described as an individual "who sits upon the back of a mule near the rear and cracks his bull whip." He described the road as "crooked as a ram's horn."

One half way from Crook to Deadwood they struck the old stage road made famous by agents and express messengers including Boone May, Scott Davis, Gale Hill, Jesse Brown and Bill Sample. Climbing a hill upon the old Sidney Road was a ranch at the junction of the gulches (which likely was the old Half-Way House) and a thousand feet below were the log cabins and buildings—a scene of many a "weird orgie that sent many a corpse to the little private cemetery back of the house," wrote Amos.

In ten minutes the Deadwood Reduction Mill ruins—destroyed by fire —was reached. "Deadwood," related the traveler, "had seven thousand people, now four thousand. Then a magician told them how to reduce refactory ore and over six thousand dollars was raised to build a mill." (The mill was rebuilt after the fire and then purchased by Golden Reward.)

At the foot of the Long Hill (Deadwood Hill) was a toll gate and one entered Chinatown. There were a few of the original cabins standing but most of them were empty. The Mongolians had several frame houses and the street was unpaved and dirty. Three minutes later they passed the turntable used by the horse cars. Deadwood was a cosmopolitan city with Asiatics, Africans, European-Oceaneans, Americans, Scots, Spanish, French, Ethiopian, English and Italians. Fruit was sold in the streets and chestnuts offered for sale at the corners. There was a good German restaurant with beer, pretzels and weiner schnitzel. The Daily Times was running on water power—at eighteen dollars per month.

Cummings learned that about 30,000,000 pounds of freight were handled annually—Wells Fargo averaged 40,000 pounds a month. There were monthly shipments of bullion, $250,000 at a crack. Six hundred and twenty-five men handled the freight. Houses rented for ten to forty dollars in the gulch city, there were five churches, three banks, telephones, water that was controlled by the Homestake Hydraulic Company. And there were tourists wandering about—they aren't a recent phenomenon by any means. Amos also visited Lead, the City of Mills, and rode on the Homestake Railroad that he termed as one of the "highest and crookedest railroads in the world."

In 1887 C. G. Grallup carried the mail from Crook City to Deadwood six times a week . . . and at the toll gate had to pay thirty-five cents. Mr. Grallup thought since he was such a regular passerby that twenty-five cents should be sufficient. Myra Ritchie the gatekeeper didn't agree and when offered the two bits refused it. The other was insistent and when he started going through the gate Ritchie took hold of the reins and backed the mail team . . . Grallup was angered and hit him with his whip. And then to top it off he filed a complaint with the United States commissioners that the toll house keeper detained the United States mail!

Boulder Canyon to the east of Deadwood was a "howling wilderness" in 1876 but by April of 1879 there was a lot of plowing in the park and people had intentions of having a townsite there the following summer. There was considerable excitement with the discovery of tin and already silver, chalk and three shades of marble had been found. A road was made through to the gulch city and the overland stages started coming through here instead of going around by Crook City. Several freighting outfits used this route, too.

The Boulder Canyon Road was filed by J. M. Gilman in 1879 and he constructed a road from Red Rock to Deadwood. He sold the Sturgis end of the road that went through Bear Butte Canyon—this stretch of road four and one-half miles from the Key City. A man named Harmon purchased it in 1881 for six thousand dollars but had a hard time proving ownership as the records burned in 1879. There was a toll gate south of the school—the old trail went south of the schoolhouse, on the southerly fringe of the park and followed a picturesque small canyon for a bit and then to Sturgis south of the present road. An old-timer remembers that there also was a toll gate at the lower end of Deadwood not far from Split Tail Gulch.

Gilman had trouble with his toll road that went through the Mountain Ranch area, west of the present Oak Ridge Cemetery—this in 1893. The First National Bank of Deadwood, owners of the land, fenced off the road with barbed wire and hauled in giant logs.

"Not fair, not fair!" protested Pete Gilman and the matter was brought to court. Said Pete, "This road has been in use for fifteen years by the traveling public. It is also used by a large number of camp workers as there are many logging operations and sawmills in the area." He also admitted the road was a profitable enterprise.

The bank people replied that they had left a good and safe roadway along the section line and had even spent one hundred and seventy-five dollars to make a more preferable road. Judge Plowman, who sat on the case, decided in favor of Gilman and the fence had to come down and the logs removed . . . and Gilman happily continued with the operation of his road.

The first road to Lead from Deadwood was over McGovern Hill, then a toll road was constructed that went up Whitewood Creek, through Cleveland where a dugout was made in the cliff (near the Claim Jumpers Inn), it went past Peck's Gardens—the area near the Northern Hills Nursing Home and Cappas's Pottery Shop—and then to Lead up a branch of Gold Run to the Washington section of Lead. At Peck's Gardens, called hot-house gardens by the early newspaper, early vegetables were grown to sell in the mining camps. In 1888 the paper mentioned that the Cleveland Toll Gate was opposite Peck's Gardens.

The road via Washington was replaced by a new toll road that went up Main Gold Run and in July of 1881 it was noted that it was "completed to the Homestake Dam about two hundred yards below the mills and the remainder of the road work should be completed in about a week." The residents of Highland Park (near the present Kenwood Apartments) had a scare in 1896 when it was suggested that the early road through there might be opened again . . . and they had it all fenced in. A reporter wrote, "These greedy people would have to take them down . . . they weren't as independent as they supposed they were."

In 1883 there was a fearful flood that did a lot of damage to all the roads in the county—it also claimed three lives at the toll gate on Whitewood where Mr. Chandler was the toll gate man. Mr. Chandler's body was swept by the flood down the creek and was found near Whitewood Falls, Mrs. Chandler was found near Fargo's Store and the body of a man who was visiting them, a Gustaf Halhuten, was found in upper Sherman. It was said that Halhuten had a considerable number of gold coins and may have been trying to rescue them when the flood waters hit.

On Wednesday, May 16, 1883 about four o'clock in the afternoon Daniel Rathburn at Ten Mile phoned to Deadwood saying the snow was leaving the high hills and Whitewood was higher than he had ever seen it. He warned the gulch people to prepare for a deluge.

And a deluge it was. Water caved in the Lee Street bridge and toppled Sam Cushman's building into the creek where it threatened to dam up the waters. A black man—labeled as a brave individual by the press—saturated it with coal oil and set it afire. The flood started washing away embankments that had been erected to beautify and protect the residences. Levees were erected in a number of places—on lower Sherman and Wall one was built with sacks of oats.

These were feeble efforts and the flood went its angry way doing an enormous amount of damage. A two-story house and stable were undermined and "turned toward each other in a loving manner." Tom Manning's livery stable tumbled in the creek—a giant mass of splinters. Another stable sailed down the stream as did the Northwestern barns. Warehouses, the seminary building, restaurants, the new school building that had cost $12,000 to build set out to sea. The Methodist Church that had been open only two months was lost, Asmon and Ayres lost their tin shop and a pawn shop "drowned." Cats, dogs, chickens, hogs, horses, cattle, pianos, organs, wagons, buggies, stages and a hearse fell in the drink . . . and a safe.

There also was extensive damage at Golden Gate and South Bend. In Galena all bridges were swept away and a brewery building used as a stable went down Bear Butte, as did several log houses. The newspaper reported that A. J. Wood's new boarding house "shook hands with Colonel Davey's bridge and they waltzed to the prairies together."

The area people agreed that the flood had done one good thing—it swept away an enormous amount of filth and debris as well as subjecting the denizens of Bull Flats to a thorough washing!

Some of the most serious damage was bourne by the toll roads. The Gayville Toll Road had damages amounting to $3,500 . . . in fact Crist said it was about ruined. Loss to Gilman's road, house and furniture was estimated to be $2,000 and Harmon and Company had $8,000 worth of damage. Dan Rathburn's toll road along Whitewood Creek from Ten Mile to Pennington was almost completely washed out and he applied to the county commissioners for financial help. They refused saying they had more roads than they could take care of. Least amount of damage was suffered by the Lead Toll Road Company although they did lose some bridges. However, before long most of the roads were repaired and open to travelers.

Bill Sheckel who grew up in Pluma remembers the toll house located near the old power plant where the road crosses Whitewood today. Colonel Thompson was the gate man—a dyed-in-the-wool Democrat. A visitor noted that the good colonel was an excellent housekeeper, in fact when making dinner he wore an apron and washed his hands! This particular time he served some feather-light biscuits, boiled ham, spuds and fish chowder—exceedingly tasty fare and his guest went "his way rejoicing."

In the early days people went swimming in the Homestake settling dam and the newspaper tsked-tsked that these frisky bathers should wear more than their birthday suits as a good many of the stage passengers going by were ladies.

Later toll roads included one to the Welcome country at the head of Terry Gulch—an early map shows the toll gate about one quarter mile down from the confluence of Willow Creek and Whitetail. There was a toll road to Custer that was apparently west of Lead, up Summit Hill where at the top there was a fork with one branch going to Ten Mile and the other to the west. It is said the toll gate was at the foot of Summit Hill.

Another later toll road was one from the Elk Creek Station to Tilford—people had donated eight hundred dollars toward its construction and then A. M. Morse charged toll. The contributors took a mighty dim view of this and Morse was persuaded to quit charging. There was a toll road to Forest City in the Tinton country and another in upper Deadwood Gulch to the mines in the Ragged Top country as already mentioned. A road, with a toll gate evidently at the mouth of Grizzly Gulch, led to the Rochford country—this road now buried under the Homestake Dam.

Not always were the roads kept in good condition and there was considerable squawking about the deplorable state they were often found. There also was a lot of grumbling about the tolls charged that varied widely. At one time it cost fifty cents to get to Central City from Deadwood and seventy-five cents for a round trip. The cost was one dollar to Lead. Some road companies charged forty cents, others thirty-five and some a quarter . . . but then the government stepped in with the Toll Road Commission that set the rates. They were established at twenty-five cents for a two-horse team and ten cents for each additional span.

Government interference plays havoc with free enterprise as the Gold Run people found out. They were facing stiff competition with people using the Poorman Road and wanted to lower their rates to entice customers—but the commission said, "No."

Whatever the fees there were individuals who complained about them. One was David Price, a hackline operator, who refused to pay toll and was hailed into court. Justice Barker fined him twenty-five dollars—that would have paid for many toll gate fees.

There is a story of a man leading a small donkey arriving at one of the toll gates and the operator came out with hand outstretched for his money. The traveler diligently searched each pocket but couldn't come up with even one cent. He searched again. Nothing. Then—an idea. "Do you charge for foot passengers?" he asked. The mystified toll keeper answered in the negative.

"Very well," said the traveler, picked up his dimunitive jack and walked through the gate toll-free.

One time a hackman named Johnson cut his rates by one-third and his rivals removed the burrs from his hack, the wheel fell off and his passengers had to walk back to town—so it wasn't only the toll men who had troubles.

The toll roads were used by the mail carriers for local mail and passenger service. In 1878 Rogers carried the mail to Bear Gulch from Deadwood. In 1884 the mail went three times a week to Vale by the way of Crook City and Butte—there were several towns of Butte but this particular one was four miles west of Whitewood on the Belle Fourche. C. G. Grallup carried mail to Crook City six times a week—this was around 1887. An important mail road was a free one, this was the Carbonate Road that started from Sawpit Gulch, near Central City, went past the Sir Roderick Dhu Mine, across Polo Creek, then along the divide of Whitney Gulch to Carbonate—there also was a road that went from that camp to Spearfish. This road was constructed with the expense bourne by Lead, Central City and the valleys with a half-way house near the present Columbia Dam. Remaining at the site are boards, part of a stove and other objects. In 1897 Eugene Walton purchased the mail contract and passenger service to the silver camp from John McLaughlin and planned to make the trip every other day—as soon as the weather moderated a bit. Mr. Walton had two choices for his route—through Blacktail (that had a toll gate) or through Sawpit Gulch. Chris Jensen carried the mail from Montana City, Deadwood, Rochford,

Lookout and Hill City and was noted as being part of every household on the route. One time he brought in some medicine for a patron on the way, at times he'd bring root beer, calico, express packages. One November day in 1889 he took a bedstead and bedding, plus a cookstove from Cleveland to Hill City. In Lead City he picked up a demijohn and jewelry and at the old toll gate on Whitetail a man handed him a trunk for someone in Hill City. A very accommodating gentleman indeed.

John McClintock had a hack between Deadwood and Golden Gate—he later had a stage to Spearfish as did Rogers and Speirs. Charles Eaves had an omnibus on the Upper Deadwood Gulch Road, Alexander a hack line to Lead, Agno a local route until he took the stage route to Rapid City, Ingall had a four horse daily to Galena . . . and there were many others.

The hacks and expressmen had a little recess about 1889 to 1893 between Lead and Deadwood. At the time a horse car ran one and one-half miles in Deadwood to connect with a narrow gauge to Lead that was owned by a New York syndicate. On the line were a little mountain engine, two light passenger cars and two flat cars. It was quite a saving to travel by rail as the charge was only twenty-five cents or forty cents for a round trip—and the trip was made every two hours.

The Dutch Count of Lead wanted to ride this new-fangled hack but the derned thing wouldn't wait for him, although he shouted, jumped and waved his arms for it to stop. A fine howdy-do fumed the Dutchman.

The Deadwood Central was abandoned in 1893 and the hacks were back in business with their, "Deadwood, D-e-a-d-w-o-o-d, going right down," or "Lead, L-e-a-d, going right up. Room for one more passenger . . . come a-runnin'."

Gradually free roads were constructed—one started on Deadwood's Main Street, went up to the head of City Creek near Mount Roosevelt (Gay's Peak of the early days), then over to Garden City where one road branched to Carbonate and the other to Spearfish, via Polo Creek. This road was described by an individual as "being utterly oblivious to the points of the compass as the lines weave in and out, curving here and there to avoid a rock knoll, there as if to leave its shadow." A good description of many of the Hills roads.

The County Commissioners offered Ellen Scott and the Deadwood and Gayville Toll Road Company one thousand dollars for their road—and which was refused. However, the board declared the road open and the sheriff delivered an order not to collect tolls. Five hundred dollars for the Blacktail and False Bottom Road was tendered and Gilman three hundred dollars for the Boulder Canyon Toll Road. In 1896 the road at Dan Lane's Sawmill in Miller and Cronan Gulches in Centennial Valley was declared open and the same year the road was opened from Portland to Long Valley and Squaw Creek. This ended the era of toll roads that had played an important part in the early transportation history—that is for twenty-five years, a quarter of a century.

The roads under the direction of the county and state gradually improved. One lanes were replaced with two lanes and then four lanes where possible. The surfaces were made all-weather. Sturdy bridges were constructed. A giant step from the days when a reporter described the Gayville Road—a carbon copy of the others—as one of pure unadulterated mud—its depth unfathomable, its color indefinable, its extent illimitable. He believed that any pedestrian who did not offer a prayer on the wayside was foolish as he observed a wayfarer lose an overshoe, noted a sunken boot and a hat lost to sight forever . . . .

*Preacher Smith who met his death on the Deadwood-Centennial Toll Road where ". . . the Hills give way to the prairies." He is buried on Mount Moriah in Deadwood.—Photo courtesy Phoebe Apperson Hearst Free Library; Meddaugh Collection.*

*This was the Nobel Toll Gate at the east end of Deadwood on the road to the north. In 1890 Dryling was the toll house keeper.—Photo courtesy Lawrence County Centennial Collection.*

*The original Thompson Half-Way House was a large building a short distance southeast of the later Half-Way House (Cliff House). Mr. and Mrs. Peter Thompson were the operators of the stop where they had a tavern, hotel and a barn. —Photo from: Sorensen, Marie Thybo "My Black Hills."*

*North of Centennial City on the stage and freight road from Crook City is this stone house with all the wooden parts burned away—likely in the fire of 1893, one of the worst in the Hills.*

*This saloon at Valley Ranche is still standing—the original logs covered over. It is located on the Duane Carstenson Ranch.*

*North of the present Frawley Ranch is this rejuvenated building that was the Dryer Ranch and a good stopping place for later day pilgrims.*

*Up Deadwood Gulch were a number of mining camps including Gayville, the first "leading city," that was laid out by William Gay on April 1, 1876 and soon had 250 homes, 30 business houses—and a brewery.—Photo from J. I. Sanford, "Black Hills Souvenir." 1902*

*The Black Hills Brewing Company up Deadwood Gulch. The brewing companies gave a lot of business to the freighters and draymen bringing in the raw materials for the brews and then distributing the finished product.—Photo courtesy Mrs. Mel Hendrickson.*

*This was the toll house in Upper Deadwood Gulch at Gayville, approximately opposite the present Bob Ruth home and below the county highway yard. In 1882 Crist moved in the toll house and operated it for many years. In 1890 the channel of the creek was moved when a grade was constructed for the railroad that went up to Lead from Deadwood.—Photo courtesy Adams Memorial Museum, Deadwood.*

*Marsh and Barnett ran the Central City Bazar. Central City was the baby of the Upper Deadwood Gulch camps of Gayville, Blacktail, South Bend, Golden Gate and Anchor City but in time became the leading city of the lot.—Photo courtesy Alumni Association, School of Mines.*

*Blacktail where there was a toll road to Garden City to the north . . . the later Maitland. The toll house is in the middle of the photo.—Photo courtesy Mrs. Helen Cooper.*

*A close-up picture of the toll house taken in 1962.—Photo courtesy Mrs. Meredith Nelson.*

*Mr. George Hunter (93 years old) on the Deadwood Hill road in east Deadwood. He remembers when there was a toll gate (Gilman's) on the other side of the gulch where the early road went to Sturgis. He also remembers Nobel's Toll gate below toward Deadwood on the righthand side of photo. Back of Mr. Hunter and below were the Deadwood and Delaware Smelter works. In 1897 the D&D Company received permission from the county to move the road further north "under the limestone cliff"; this was necessary after the slag reached the road.*

*An early photo of Deadwood Gulch where gold was discovered in the early days.—Photo: Black Hills Illustrated.*

*Photo taken in Oak Ridge Cemetery east of Deadwood. Looking to the west was Mountain Ranch on the other side of the fence; from here the early road went to the south, crossing the present highway and then turned westerly toward the gulch city.*

*In Boulder Canyon was this interesting cabin of an early day settler who raised chickens . . . the house was torn down when the modern road was constructed. In 1880 it was noted that the Boulder Road was well graded and good for traveling—usually there was considerable complaining about the roads so this was quite a switch.—Photo courtesy Genevieve Lorenzo.*

*A toll road went to Terry located west of Lead in area that first came into prominence due to silver discoveries. The toll road was operated by the Lead Toll Road Company—a hack man on this route was Art Taylor shown in the photo. Fred Doten had a hack line that was purchased in 1895 by Richard Harris. The old Pierre coach—a tallyho—was used between Lead and Terry.—Photo courtesy Ed Johnson.*

*Moses Manuel one of the discoverers of the Homestake Mine that launched the Homestake Mining Company into being one of the greatest gold producers.—Photo courtesy Alumni Association, School of Mines.*

*The Homestake Stamp Mill and wood chutes in Lead—part of the operations of the mining company.—Photo courtesy Phoebe Apperson Hearst Free Library; Meddaugh Collection.*

*Mill Street in Lead . . . the south part of the road led to the Pennington Road that went down to the present Kirk where Northwestern kept their mules, horses, where they maintained a blacksmith shop and a round house for their employees.—Meddaugh, Souvenir of Lead 1892.*

*North of Deadwood on Centennial Prairie Michael Burton kept horses for Deadwood residents—at a cost of two dollars per month plus delivery. This photo shows the trench of the stockade that was on a small knoll that provided a clear view in all directions. Elkhorn Mountain is in the back.*

*The spring house north of the Burton Stockade on the present day Glenn Baker Ranch. The roof has been replaced but the rest of the building is believed to be original.*

*Early mail service was to Minnesela, east of the present-day Belle Fourche. This is the old post office building on the ranch of Mr. and Mrs. Earl Gaver.*

*Minnesela. The town gave way to Belle Fourche when the railroad went to the latter town. In 1893 a young horse thief named Ducharme was jailed in Minnesela—he and his gang had appropriated stock west of the cheese factory on Horse Creek.—Photo courtesy Clyde Mitchell.*

*The opposite direction from Minnesela was Rapid City—here is a photo of the place in its salad days.—Photo: Souvenir of Grand Lodge IOOF 1909*

# BIBLIOGRAPHY

Andreas Historical Atlas of Dakota, D. T. Andreas Publishing, Chicago 1884

Best, Jack Carson *The Unwickedest Outlaw,* 1966 C. Phil Cooke, The Press of the Territorian, Santa Fe

Black Hills Times 1877-1897

Black Hills Pioneer 1876-1897

Brevets *North Dakota Historical Markers of Sites,* Brevet Press, Inc., Sioux Falls, South Dakota 1975

Briggs, Harold E. *The Black Hills Gold Rush,* North Dakota Historical Quarterly, Volume V, No. 2, January 1931

Briggs, Harold E. *Early Freight and Stage Lines,* North Dakota Historical Quarterly, Volume III, No. 4, July 1929

Brown, James, Letters, privately printed.

Brown, Jesse and Willard, A. M. *The Black Hills Trails,* Rapid City Journal Company, 1924

Brown, Mark H. *The Plainsmen of the Yellowstone,* G. P. Putnam's Son, Inc., New York 1961

Clark, Badger *When Hot Springs Was a Pup,* Lame Johnny Press Association 1976

Clements, Ahme, Medora-Black Hills Stage Line, Collections of the State Historical Collections, Volume 25 South Dakota

Crawford, Lewis F. *The Medora-Black Hills Stage Line,* Collections of the State Historical Society of North Dakota, Vol. 7, 1925

Curley, Edwin *Curley's Guide to the Black Hills,* Facsimile Dakota Wesleyan University Press, Mitchell, South Dakota 1973.

Dakota Weekly Register, Spearfish. June 4, 1881. August 1883.

Deadwood Pioneer Times 1897-1900

Echoes Through the Valleys, Meade County History

Hamilton, Will *An Autiobiography of a Cowman,* South Dakota Historical Collections

Hanson, Myrle G. *A History of Harding County to 1925,* Thesis 1925

Hebert, Frank *Forty Years Prospecting and Mining in the Black Hills,* Rapid City Journal

Heidepriem, Eric *Original Trail Retraced,* Rapid City Journal, May 7, 1972

Hillinger, Charles *Salesman Makes a Ghost Town North Dakota's Biggest Attraction,* Los Angeles Times

Jennewein, J. Leonard *Ben Ash and the Trail Blazers,* South Dakota Historical Collections, Volume XXV 1951

Jensen, Mrs. Pete, Powderville History

Kingsbury, George W. *History of Dakota Territory,* Volumes I & II ed George Martin Smith, S. J. Clark Publishing Co. 1915

Koury, Michael *Military Posts of Montana,* The Old Army Press, Bellevue 1970

Lee, Bob and Williams, Dick *The Last Grass Frontier,* 1964

Lemmon, Ed *Boss Cowman,* University of Nebraska Press 1969

Loss, William E. *A History of Steamboating on the Upper Missouri River,* University of Nebraska Press, Lincoln 1962

Mandat-Grancy, Edmond *Cowboys and Colonels,* (1883) J. B. Lippincott Co., Philadelphia and New York 1963

Minnilusa Historical Association

Mahnken, Norbert R. *The Sidney Black Hills Trail,* Nebraska History, Volume 30, No. 3, September 1949

Morrell, Warren, Through the Hills, Rapid City Journal 1952

McClintock, J. S. *Pioneer Days in the Black Hills,* 1939

Norton, Sterling & Watkins, Albert *History of Nebraska,* Western Publishing and Engraving Co., Lincoln 1918

Osburn, James D. *A History of the Deadwood Trail,* State University, South Dakota 1962

Palais, Hyman *A Study of the Trails to the Black Hills Gold Fields,* Volume 25, South Dakota Historical Collections, Vol. XXV 1951

Pratt, Alice Day *Three Frontiers,* Vantage Press, Inc., New York 1955

Shevling, Sara, Diary, Unpublished.

Sidney Telegraph, January 6, 1877

Somner, Lawrence J. *Fort Pierre, South Dakota. A Preliminary Assessment of Historic Preservation Opportunities,* Sponsored by Community Development, Fort Pierre, SD, January 1976

South Dakota State Historical Resource Center, Department of Education and Cultural Affairs, Pierre, South Dakota

Sorensen, Marie Thybo *My Black Hills,* privately printed.

Souvenir of Grand Lodge, IOOF 1909

Spring, Agnes Wright *The Cheyenne and Black Hills Stage and Express Routes,* Glendale, California, Arthur H. Clark Co. 1948

Stephens, Grace, Hough Family History, unpublished

Sutley, Zack *The Last Frontier,* The Maxmillian Co., New York 1939

Szalay, Eugene, Notes on Miles City route

Tallent, Annie D. *The Black Hills or Last Hunting Ground of the*

*Dakotahs,* Nixon-Jones, St. Louis 1899

Tidball, Rose *Taming the Plains,* 1976

Vanishing Trails Expedition, The Vanishing Trail Committee, Compiled and Edited by Dean S. Nauman 1976

WI-IYOHI, Bulletin of South Dakota Historical Society. Diary Frost, W. H., XIX, April 1965

Watson, Parker *Gold in the Black Hills,* University of Oklahoma Press, Norman, Oklahoma

Wolle, Muriel Sibell, *The Bonanza Trail Ghost Towns and Mining Camps of the West,* Indiana University Press 1953

Wright, Dana *The Black Hills Trail from Fort Abraham Lincoln to the South Dakota Line.* North Dakota History, Vol. XXV-1, Jan. 1958

Wright, Kathryn *The Real Calamity Jane,* True West, December 1957

Yellowstone Journal October 9, 1879

## INDEX